THE ARTIST
& THE SCIENTISTS

The Artist and the Scientists: Bringing Prehistory to Life chronicles the extraordinary work of pre-eminent palaeontologists Patricia Vickers-Rich and Thomas Rich, and Peter Trusler, arguably one of Australia's — if not the world's — most accomplished artists of scientific realism.

Over more than thirty years Pat, Tom and Peter have travelled together, across Eastern Europe, Asia, the Americas, Africa, Australia and New Zealand in search of the remains of early life, including fish, dinosaurs, birds and mammals. Their successful expeditions, and the many publications and exquisite artworks that have ensued, are reflected upon from their unique and individual perspectives, capturing both their scientific rigour and immense camaraderie.

With over 200 illustrations comprising stunning reconstruction art pieces, from preliminary sketch to finished form — many never before published, as well as photos of the sites and people involved, this book covers the last 600 million years of the geological record. *The Artist and the Scientists: Bringing Prehistory to Life* will appeal to a huge readership, from amateur dinosaur enthusiast, to evolutionary scientist and reconstruction artist alike.

Peter Trusler is a freelance palaeo-reconstruction and biological artist, and is a zoology science graduate from Monash University.

Patricia Vickers-Rich is Professor of Palaeontology and Founding Director of the Monash Science Centre at Monash University.

Thomas H. Rich is Senior Curator of Vertebrate Palaeontology at Museum Victoria, where he has worked for 36 years.

THE ARTIST & THE SCIENTISTS

BRINGING PREHISTORY TO LIFE

Peter Trusler, Patricia Vickers-Rich, Thomas H. Rich

CAMBRIDGE UNIVERSITY PRESS
Cambridge, New York, Melbourne, Madrid, Cape Town, Singapore, São Paulo, Delhi, Dubai, Tokyo, Mexico City

Cambridge University Press
477 Williamstown Road, Port Melbourne, VIC 3207, Australia

Published in the United States of America by Cambridge University Press, New York

www.cambridge.org
Information on this title: www.cambridge.org/9780521162999

© Peter Trusler, Patricia Vickers-Rich, Thomas H. Rich 2010
Artwork © Peter Trusler (unless otherwise indicated)

This publication is copyright. Subject to statutory exception and to the provisions of relevant collective licensing agreements, no reproduction of any part may take place without the written permission of Cambridge University Press.

First published 2010

Designed and typeset by Sardine Design
Printed in China by Printplus

A catalogue record for this publication is available from the British Library
National Library of Australia Cataloguing in Publication data

> Trusler, Peter.
> The artist and the scientists : bringing prehistory to life / Peter Trusler, Patricia
> Vickers-Rich, Thomas H. Rich.
> ISBN 9780521162999 (pbk.)
> Includes index.
> Bibliography.
> Arts and science.
> Paleontology.
> Paleontologists—Australia—Biography.
> Artists—Australia—Biography.
> Rich, Patricia Vickers
> Rich, Thomas H. V.

560

ISBN 978-0-521-16299-9 Paperback

Reproduction and communication for educational purposes

The Australian Copyright Act 1968 (the Act) allows a maximum of one chapter or 10% of the pages of this work, whichever is the greater, to be reproduced and/or communicated by any educational institution for its educational purposes provided that the educational institution (or the body that administers it) has given a remuneration notice to Copyright Agency Limited (CAL) under the Act.

For details of the CAL licence for educational institutions contact:
Copyright Agency Limited
Level 15, 233 Castlereagh Street
Sydney NSW 2000
Telephone: (02) 9394 7600
Facsimile: (02) 9394 7601
E-mail: info@copyright.com.au

Reproduction and communication for other purposes

Except as permitted under the Act (for example a fair dealing for the purposes of study, research, criticism or review) no part of this publication may be reproduced, stored in a retrieval system, communicated or transmitted in any form or by any means without prior written permission. All inquiries should be made to the publisher at the address above.

Cambridge University Press has no responsibility for the persistence or accuracy of URLs for external or third-party internet websites referred to in this publication and does not guarantee that any content on such websites is, or will remain, accurate or appropriate.

Foreword

Confessions of a Dinophile

'The roof has fallen in on me in the last 24 hours – the most ambitious book project of my life – my first non-SF novel! I've had to cancel everything, including a major TV program. Nothing more I can do on *Dinophile*. I suppose you've seen the story in today's papers about the discovery in China of a four-winged dinosaur! Right up your street. All best, Arthur 24 Jan 2003'

[Pat: So, with that pressure on, Arthur still put forward the foreword for a book that he, Tom and I had discussed with him during our 2003 visit to his home in Colombo. It went something like this.]

My fascination with ancient things, especially dinosaurs, started earlier than most people's – when I was about 10 years old. I can still see that first awakening – though it may be a false memory. I was sitting beside my father not far from our home in Somerset, England – in an open pony cart. He had just opened his cigarette pack, and stuffed inside it was a little card. That card had on it a weird monster with a row of plates sticking up from its back – later I found out this was a stegosaur. The purpose of those bony features on its back is still being debated: perhaps they were for collecting (or re-radiating) the warmth of the sun.

Why the cigarette makers in the 1930s were so taken with dinosaurs is puzzling – maybe their public relations head was fascinated, like me, with these strange and wonderful beasts, but nonetheless another company picked up the interest and put out a beautiful collection of 3D photographs – so life-like that they could have been real. I had the collection for many years, and hope it is still lurking somewhere in the 'Clarkives'. So ironic that I owe my early interest in science to the drug that killed my

Arthur C. Clarke (centre) with Pat and Tom during their visit to Sri Lanka in 2003 (photo courtesy of Arthur C. Clarke).

father, but thankful that I did not take up the habit. And, I have to say, it was the impact of this early acquaintance with something scientific that made a difference in what I decided to do later.

Another powerful reinforcement of my interest in the past was the 1925 version of Conan Doyle's *The Lost World* – the first movie I can ever remember seeing. I can still recall scenes from it, particularly when the dinosaur gate-crashes a typical London club and wakes up even its most somnolent members. I must also have seen the later 1960s version, but I can't remember anything about it – those first impressions stuck.

Many years later, I had another rather surprising encounter with things ancient – an encounter that dealt with the demise of a very successful group, the dinosaurs, under mysterious circumstances. In 1942, Flight Lieutenant Clarke took charge of the experimental Ground Controlled Radar developed by an MIT team led by Luis Alvarez (hence my novel *Glide Path*, which is dedicated to Luis and my colleagues in the US Air Force and the RAF). Well, in 1980, Luis and his geologist son Walter advanced the theory that the dinosaur extinction circa 65 million years ago was caused by the impact of an asteroid or comet, with catastrophic consequences to life on Earth. Other mass extinctions before and after that of the dinosaurs have also been attributed to such extraterrestrial invaders, even during the most ancient of times. There is now no doubt that the Earth has been battered many times in the past. The face of the Moon is clear evidence of what can happen to a world unprotected by an atmosphere!

To me, it was the images of the past that introduced me to something other than today. The amazing success of Steven Spielberg's *Jurassic Park* series proves not only the popularity of one ancient group, the dinosaurs, but also the impact of high quality imagery. One day, perhaps, dinosaurs may return – through cloning, if viable DNA samples can be discovered, or more probably through computer regeneration! Then *Jurassic Park* would become a reality. But for now we must depend on reconstructions, either in our own mind or given to us by those who have the knowledge combined with the artistic skills which can allow, almost magically, a rendering of these ghostly images with brush and pen or increasingly on computer screen! But computer images still need the human brain to construct them – now, but perhaps not in the future?

The Artist and the Scientists is a book that delves into the detail of how such palaeo-reconstructions of times past are put together. It is unique, for it has been put together by a fine artist, Peter Trusler, who revels in detail and is a stickler for checking out everything himself (not taking in anything without evaluating it thoroughly), and two scientists, Tom Rich and Patricia Vickers-Rich. The trio has worked together for three decades. Like Peter, both researchers are willing to go to extremes to tease out the scientific detail and both have been determined to make sure there is funding to ensure that the time-consuming attention to detail can be underwritten – not an easy task in ever more challenging economic times. This book is many of their stories, centred around several reconstructions that they put together, together. The book is not a history of scientific art, nor is it an evaluation of other artists' and scientists' renderings. This book is a personal account of the interaction of an artist and two scientists, describing how they went about giving us their marvellous insights into the past – a bit of a step up from those collector's cards that my father gave to me so long ago!

Arthur C. Clarke, 2003

'*Leptoceratopsian*' (1999). Watercolour and gouache on paper, 20 x 30 cm, private collection (P. Trusler). *Serendipaceratops arthurcclarkei*, a frilled dinosaur from the Early Cretaceous of Victoria, was named in honour of noted science fiction writer Arthur C. Clarke when Pat and Tom discovered that Clarke's early interest in dinosaurs was what enticed him into science. The reconstruction is based on the better-known *Leptoceratops gracilis*, the species most similar to *S. arthurcclarkei*. The total length of the dinosaur is about 1 metre.

Contents

Foreword by Arthur C. Clarke — v

Acknowledgements — xi

Geologic Time — xii

Introduction — 3

1. A New Paint Box — 7
2. The Dragon Chasers — 13
3. The Beginning: Bacchus Marsh *Diprotodon* — 23
4. The Art of *The Fossil Book* — 45
5. Dinosaurs from China — 63
6. *Wildlife of Gondwana* — 77
7. The Dinosaurs of Darkness — 105
8. A Moa Mummy: A classic dissection — 147
9. Magnificent Mihirungs — 161
10. Where the Wind Bites: Patagonia — 181
11. *The Rise of Animals*: Back to the Precambrian — 199
12. Chinese and Australian Mesozoic Mammals — 233
13. The Last of the Mob — 255
14. The Art of Humour and Enchantment — 283

Conclusions — 293

Glossary — 297

References — 300

Index — 304

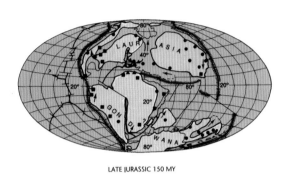
LATE JURASSIC 150 MY

LATE CRETACEOUS 65 MY

OLIGOCENE 25 MY

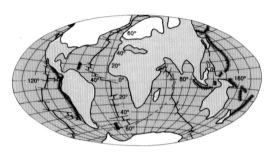
LAST GLACIAL MAXIMUM 18000 Y

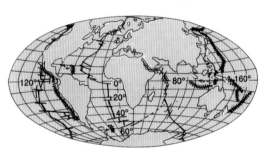
PRESENT

Maps illustrating where Australia has been over the last 150 million years (from *Wildlife of Gondwana*).

Acknowledgements

Thanks are due to so many people who have assisted each of us. The help and inspiration that we have each received during our formative years, right down to the assistance with the matters of yesterday, have all had an ultimate bearing on this book. Many have been mentioned during our three-way account of the projects herein. There are others, of course, who have provided valuable guidance and facilitated our work and interests in diverse ways. There have been those who have directly or indirectly provided mentorship and dear friends who have offered kindness and candour at pivotal times. Colleagues, technicians and professionals of every description have willingly lent their expertise. If we formally acknowledged these contributions great and small, it would easily constitute another chapter in this book. We trust that their pride in their work and their interest in ours has been of mutual benefit, because we have certainly been so enlivened by the exchange.

Some, however, stand out. Patricia Komarower has been our chief editor and research assistant, a title which belies the graciousness with which she has crafted the manuscript, proofread and organised the massive illustration file, as well as undertaken a multitude of tasks leading to the final production. Draga Gelt has courageously dealt with graphics, image recovery and all things digital. Reuben Trusler, Steve Morton, Frank Coffa, Rodney Start and Rodney Armstrong have given invaluable IT support, photographic services and image processing expertise. Mary Walters has for many years not only provided technical assistance but also proofread mountains of manuscript. Joy Evans of Johanna Cottages in the Otway Ranges of southern Victoria gave us the solitude that made it possible to write a substantial part of this book. William Birch, Kat Pawley and Rolf Schmidt assisted with the glossary. Alan Curl and his family have enabled Peter to travel much of Australia and provided such friendly patronage. Sally Simpson has accommodated Peter in the beautiful environment of the studio that has been his creative sanctuary.

Several funding sources made the rendering of the art and the research possible – the Australia China Council, Australia Post, the Australian Academy of Science, the Australian Research Council, Monash University, Museum Victoria, the Queen Victoria Museum and Art Gallery (Launceston, Tasmania), the National Geographic Society, Qantas Airways, Atlas Copco and Andrew Isles Natural History Books. Staff at the Queen Victoria Museum have provided us with high quality images of Trusler art held in their collections. UNESCO, under the auspices of the International Geological Correlation Program (specifically IGCP493), has granted funds over the years to allow each of us to conduct fieldwork and laboratory research in institutions around the world. The National Geographic Society and the Australian Research Council have long supported our fieldwork. Australia Post has kindly given permission to reproduce the images commissioned by them, not only for this publication but also for numerous educational and scientific purposes. Every effort has been made to identify and acknowledge the sources of the many illustrations used in the book, but for a few of the older photographs in particular it proved impossible to identify the photographer. We thank all those who allowed us to reproduce these images and apologise for any omissions.

We also want to express a sense of pride and deep gratitude to our families, who have lived with our intensity for decades, who have understood our passionate commitments to our work and coped with our absences and with all the difficulties that this project has brought their way. The miracle is that they are still with us!

Geologic Time

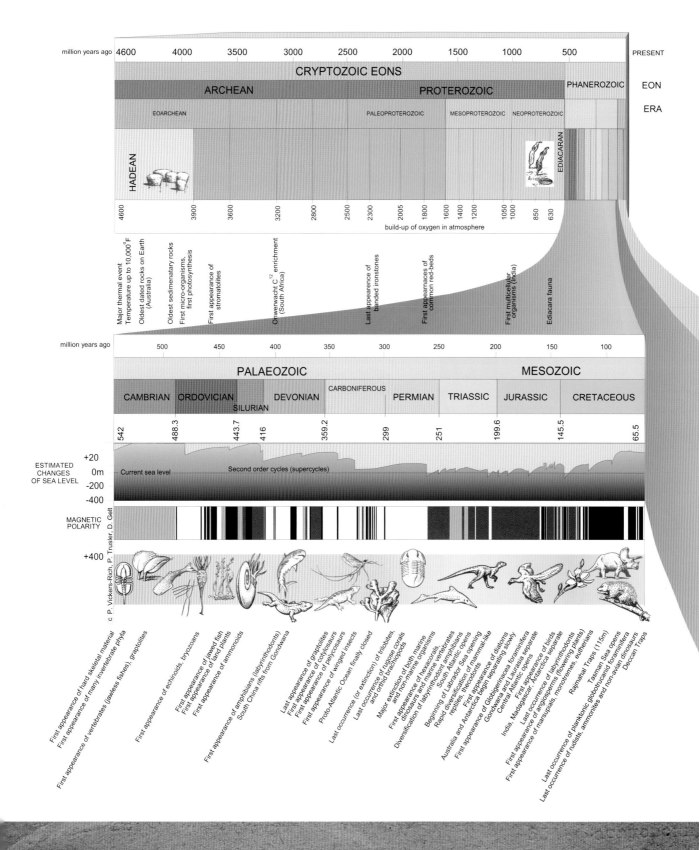

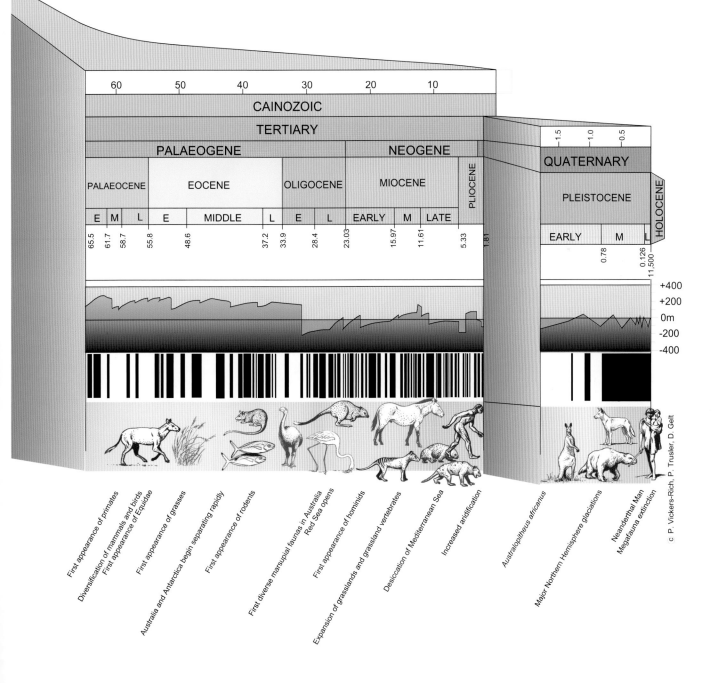

Introduction

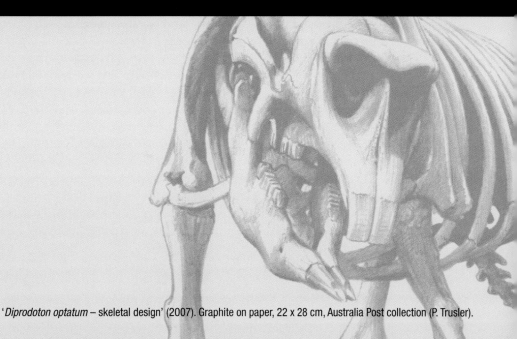

'*Diprodoton optatum* – skeletal design' (2007). Graphite on paper, 22 x 28 cm, Australia Post collection (P. Trusler).

Popularity and acceptability of art styles change over time. Some styles appeal to certain groups and cultures, at different times and in different places. Scientific art is not immune to this: if one peruses, for example, dinosaur art over the past century, significant stylistic differences are quite evident.

So, one could ask, with all the literature that has been printed over this time dealing with palaeo-art, what does one more book have to offer that has not already been covered elsewhere? What makes this book any different from all the others?

First, it considers in detail the interactions of the artist and the scientists that were vital to the production of these works of art. The description of such interaction in producing scientific art has, to our knowledge, never been done before in depth.

Another answer to the question is that this book is specific in the images it presents. And it deals with three people only – an artist and two scientists – who have worked with one another for more than three decades. That by itself would have made this book unique! But this book is different in other ways as well. *The Artist and the Scientists* considers how this close association of three people over many years allowed the exploration of the why's and how's of the production of unique art, with attention to minute detail and based on a significant body of research. Examples cover a broad range of approaches, dictated by the projects and the subject matter selected.

The book could have been presented in taxonomic arrangement: thus, fish would have come before dinosaurs, and the dinosaurs in turn before Tasmanian tigers. This has been a common approach. The book could have been divided into time periods: it could have first covered the Precambrian, followed by subdivisions of the Phanerozoic, the last 542 million years (*see* Timeline of Geologic Time). However, we decided the organisation should be based around the development of the scientific art over 30+ years, painting by painting and sketch by sketch, in order to capture how the methodology of both art and science developed over that period.

The art rendered over time also served many different purposes. The book explores those differences, detailing how particular projects were chosen, how they were managed and funded, how they were used both for scientific research and for public exposure, for educating and for entertaining in some instances. This process is complex in itself, and the dogged determination required to produce an image as close to what was the scientific reality becomes clear. Both artist and scientists were driven to seek out the detail, discuss it in depth not only among themselves but also with a wide body of research colleagues, and to take all that into account when the final plan was emplaced. This was often a lengthy process, and had to be funded, had to meet deadlines, had to avoid the final image being shortchanged because of the pressures of time and a variety of distractions imposed by life.

We, as artist and scientists, hope that this book will inspire readers, both scientific and general, to become more visually tuned in to the content of scientific art, which serves to pull together a multitude of threads resulting from research. We hope that the art in this book will communicate with and entice the more general reader into exploring the complexities of science – discovering, bit by bit, the fine detail after having initially been attracted to the bigger picture. To assist in this process, an explanation of selected scientific and technical terms has been included in a glossary at the back of the book.

Much of the art in this book has already served as a portal stimulating scientific discourse, sometimes challenging prevailing paradigms and sparking debate – which is the core of science. Some of the art, using colourful and imaginative style yet rendered with respect for scientific data, has also enticed young people into thinking about their past. And now, with the variety of styles presented in this book, and the explanation of how and why it was developed, we all hope it will encourage an attention to scientific detail in reconstruction art in the future.

The authors and a friend: from left to right, Peter, Tom and Pat with fellow palaeontologist Richard Fortey (photo by J. Fortey).

1
A New Paint Box

Figure 1.1 Pardalote, one of Peter's many modern bird images, which initially attracted Tom and Pat to Peter's realism.

The Artist: Peter

'Introspection' is probably the most apt description of my childhood realm. I seem to have been a quiet kid who was totally preoccupied with watching and thinking; probably in a very confused way! I do not know from whence this sense of wonder came, but I managed to find an outlet and further stimulus through drawing. It was all I apparently did; the physical rough and tumble of boyhood was not my desire for the most part. My parents expressed some concern about this. Over the years I can recall tentative comments and gentle coercions that they made in this regard.

My parents had no direct interest or experience in either biology or art and may have had some difficulty in satisfying my insular pursuits. My father was a keen amateur photographer, however, and as an engineer was also an efficient draftsman. His enthusiasm for knowledge was infectious. My lasting recollection is that both parents fostered my pursuits with some pride.

Two pivotal events are clear to me. The first was a Christmas gift box of 72 different coloured pencils. I was five years old at the time and the set of Derwent brand pencils were sharpened to the verge of extinction in less than three years! The next revelation was actually due to a mistake on my mother's part. I simply longed for a box of real paints – not those little kiddy ones in pans but real ones in tubes. For my eighth birthday a set arrived and by the end of that day I had made a disastrous mess. They just did not work! When my father walked in the door at day's end, he consoled my fallen pride when he picked up the box and said: 'Oh, no. You need to mix these with turps, not water.' What a revelation this was to me! So began my passion for oil painting, and this box of paints was soon to be followed by private tuition with a local artist, Jessie Merritt. Her influence on my formative practice was to be profound. It was not long before the entire experience of painting – the paraphernalia and equipment, the multiplicity of possible techniques and even the smell of it – was captivating. My initial allergic reaction to the solvents was eventually conquered!

My time was divided between my art pursuits and my interest in birds and animals. Like most children, this was easily fostered through keeping household pets – budgies, cats, tortoises, a real menagerie of odd things at odd times. I amassed collections of feathers, bones and skulls, rocks, insects and picture cards. Through my secondary school years I became involved in the local field naturalists club, and my combined interests flourished. What was interesting about the Ballarat Field Naturalists Club was that it was made up of mainly retired people, and so the very few younger members received generous mentoring. These older members obviously imparted considerable depth and wisdom with their knowledge. This seems to have been fundamental in a

Figure 1.2 Peter at work on the Australia Post megafauna stamps in his studio, 2008 (photo by G. Narbonne).

couple of ways. Firstly, it was different to my contemporary experience of both family life and peer group: it was 'old world', and it still bemuses me that I enjoyed it, because I feel that most children would not. Secondly, it was a learning environment with unbridled information that came not via curriculum, but via personality. Knowledge came passionately, imparted by unusual people from all walks of life. My art teacher, too, fitted this profile – being in her late 70s, with a gentle nature that did not disguise her sharp intellect and conservative attitude to life. She was able to impart her broad and deep knowledge because of her very personality. Her individuality and passion were what challenged me to learn and to think about things for myself.

My formal education throughout the 1960s and 1970s was rich, and this time was characterised by a transition from traditional methods and subjects to the incorporation of many innovative learning programs. There was a paradigm shift in educational policy throughout this period. It was so profound that for a brief moment tertiary education was offered for free. Reflecting on this period of time has made me realise that in many ways the absence of constraints that I enjoyed during childhood actually extended into my early adulthood. My student colleagues and I were totally free to pursue interests without fear or favour. This allowed for an intellectual freedom that favoured learning over vocational education. These unpragmatic times are now gone.

My education in science underpinned my interest in natural history, but for me dinosaurs have never loomed large. This may surprise students of palaeontology, for I am keenly aware that for many the subject is an all-consuming passion. I can look in wonder at an ancient jade carving of a mythical dragon just as easily as I can marvel at an enormous Mesozoic bone! I have a feeling that this difference in my focus is centred in my fundamental interest in art. I am just as fascinated by the natural world as I am by the social and internal workings of the human realm, and so I find that I am often at odds with enthusiasts in all fields, art or science. There is a continuum and breadth to my interests, and I am not likely to adhere to boundaries in any sense.

Figure 1.3 Peter at Mistaken Point, Newfoundland, sketching in 2005 (photo by P. Strother).

This is not to say that I lack discrimination, for I can and do make deeply critical decisions and distinctions in both respects – artistically and scientifically. Whether I am working on a person's portrait at one moment or organising a tectonic plate diagram on the computer the next, there is a nexus of criteria I can apply to my approach in either case. One might be rooted in a Western cultural history where the psyche is able to run free, and the other might be bound by the logic of scientific discipline where my role is totally subservient to information content and the creative thought of others. Either extreme can be as simple and direct as I can make it or as complex as need be. Essentially, there is great beauty and joy to be found in all, and so any hierarchy, as profound as it may be, is somewhat artificial to me. The doing of the work is fundamental – living the moment! The high points might constitute art, independent of intention or purpose and irrespective of subject. The high points might also provide a genuine contribution to science. I don't know!

The multitude of variations and subjects in between these examples is of particular interest to me: I can dwell on the edge by sliding the aesthetic component of the art in any direction. You will find considerable variation in the work featured throughout these pages, which in turn represents a small, yet significant, part of my output. These works have been produced for, and generated by, science. The manner in which Pat, Tom and I have collaborated has allowed our work to transcend boundaries, since we value communication above all. My colleagues have appreciated not only my pictorial skills, but also the intellectual approach to my craft. This has led to a genuine fusion of our ideas and has given me the privilege of contributing as a scientist to the research process. I have personally valued our egalitarian relationship, and feel that it has extended each of us.

2
The Dragon Chasers

Figure 2.1 Underground 'mining' at Dinosaur Cove (photo by Peter Menzel).

The Scientist: Tom

Southern California, Christmas Day, 1953: I received a copy of *All about Dinosaurs* for Christmas. This was when I first learned that a person who studies fossils is called a palaeontologist. I decided to become one.

The last chapter of the book was called 'Death of the Dinosaurs'. Above the chapter title was an illustration of two tiny mammals attacking, and clearly enjoying the taste of, dinosaur eggs. 'Those', my 12-year-old mind thought, 'were the really interesting animals, our ancestors when the dinosaurs were around.' I kept my interest in those early mammals but could find out little about them.

Pat had a very different background from mine. Both of my parents had received university degrees, and I had read rather widely, including classical literature across many disciplines. Pat grew up on a series of farms in California, and her rural upbringing made her a fan of country and western music, as well as giving her the knowhow to fix a tractor and raise pigs. Whereas I talked to my dad about engineering

Figure 2.2 Tom with his mother Alberta, and his sisters Penelope (left) and Patsy, around 1945.

schemes and my mother about Chinese language, Pat was more familiar with how to gut a fish or swiftly bring down a deer with her Remington 306. We did have something in common, however. We were both fascinated by science. She began collecting grasshoppers and bees and keeping them in careful rows in her dolls' house when she was four years old, and later took out top honours at the National Science Fair in Seattle, Washington, with her science project on radiocarbon dating, after visiting the inventor of the technique, Willard Libby, when she was a junior in high school (Year 11).

Three years after I decided, at 12 years old, that I wanted to be a palaeontologist, the Los Angeles County Museum offered a course in vertebrate palaeontology for year 10 students in high school. I decided to take it. This course reinforced my determination to become a vertebrate palaeontologist, despite being told by a well-meaning curator in that field at that museum that I would never get a job as one! I looked him straight in the eye and thought to myself, 'You got a job, did you not? How did you do it? You got it because you were trained and ready when the opportunity came.' I decided that even if I eventually wound up selling shoes in Mississippi, I would still study vertebrate palaeontology. That's what I wanted to do, and I would do it, no matter what the outcome!

In 1960 I enrolled at the University of California at Berkeley as a first-year palaeontology and physics student. The introductory palaeontology course was taught by Prof. Ruben A. Stirton. In the last two lectures, he told the class about his ground-breaking fieldwork in far off, exotic Australia. I became totally fascinated with Australia,

Figure 2.3 Pat's grandparents, Kelsey (left) and Mary Belle (née Graham) Vickers at their first home in California; Kelsey was part Cherokee and Mary Belle's background was English.
Figure 2.4 Pat at the age of three, examining the sands while her parents fished; she was already interested in sediments.

nearly obsessed. There appeared to be so much more to be learned about the fossils there. Near where I lived, at about the same time, an old movie house was showing *The Back of Beyond*, a film starring Tom Kruse as a mail truck driver on the Birdsville Track, in outback Australia. I went to see it three or four times in a single week and gained my first insights about what the country was like where Prof. Stirton worked.

Pat enrolled at UC Berkeley in 1962. Lucky enough to gain a scholarship, she was the first in her enormous family of uncles, aunts, first and second cousins, etc. (her father was one of 13) to attend college. In order for the finances to work for the family, even with the scholarship, her mother and dad sold up their family home and moved with her from central California to the San Francisco Bay area, so that they could all work and live in the same rental home. She was rather overwhelmed by the size of the University of California, and so looked about for the smallest department she could find. The Department of Paleontology, with only 10 undergraduate students, and little more than 30 graduates, seemed just right. After all, she was interested in fossils and geologic time. Soon after, she was able to secure a part-time job in the department, working on a fossil collection that J. Wyatt Durham had recently collected in the Galapagos Islands. The department atmosphere was ideal, for it meant that undergraduates and graduate students were treated much the same, and there was significant time spent with the lecturers.

I was one of those graduate students, and in my sixth year at UC Berkeley I met Pat. She was an undergraduate, and I was working on my master's degree. She was working on North American fossil vultures with old world affinities, and I on some primitive mammals from the Paris Basin. We found some common ground and were married on 3 September 1966.

At the end of our degrees at Berkeley, we had to make a choice. I had an opportunity to work with Malcolm McKenna of the Frick Lab at the American Museum of Natural History on a project close to my heart, Mesozoic mammals from a site in Wyoming called Como Bluff. Pat had begun working on the Australian dromornithid birds that Prof. Stirton had collected (*see* Chapter 9), and she was to be allowed to take those with her wherever we went. We both wished to go to a different university to gain the experience of another research group, and we were encouraged to do so by the staff we so admired and had worked so well with at UC Berkeley. Most of them had followed the same path, getting their educations at two or more institutions, thus broadening their backgrounds and being exposed to different intellectual environments.

Pat considered the University of Florida, for the well-known palaeo-ornithologist Pierce Brodkorb was there. At New York's Columbia University, however, there was the opportunity not only for me to work with Malcolm McKenna but also for Pat to work with Walter Bock, a well-known evolutionary biologist and functional morphologist. With that combination and the offer of two full scholarships, the choice was made.

So in late 1967 we began our studies at Columbia and the American Museum of Natural History. It was at an auspicious time: the great debates concerning stable and dynamic continents were raging, and our graduate work was being carried out amidst this. These scientific debates were focused at the Lamont-Doherty Geophysical

Labs, located up the Hudson River from both Columbia and the American Museum. Researchers, including the graduate students, from all three institutions regularly got together to discuss the issues of spreading ridges, plunging crustal plates and biogeography. Marie Tharp and Bruce Heezen, working with Maurice Ewing, were mapping the sea floor of the North Atlantic, while Lynn Sykes and Xavier Le Pichon were trying to describe what was going on below the sea floor. And we were trying to work out how our animals had managed to move about on these ever-changing bits of continents. It was a formative time for us: watching scientific debate in action; watching old paradigms being challenged – some remaining robust, others changing to respond to the new data coming in fast and furiously from the DSDP (Deep Sea Drilling Project), various seismic and regional surveys. Masses of data were being collated at a few principal institutions around the world, including Lamont-Doherty. It was an exciting time to be a graduate student, to be part of one of the international think tanks where our input was valued. Pat's work in particular, dealing with the origin and evolution of the Australasian avifauna, was central to some of the ongoing debates. We had definitely landed at the right place at the right time for our graduate studies.

During our time at Columbia, Pat and I had the opportunity to take part in two mind-expanding expeditions, both relating well to 'bumping up' our skills in our attempts at interpreting the past. Pat spent two months in Costa Rica involved in modern ecological studies, and I went to Antarctica to take part in a high-latitude expedition gathering information on Mesozoic vertebrate faunas.

Somewhat later, in 1971, Pat and I were invited by Richard Tedford to be part of the American Museum of Natural History's field expedition to explore the Tertiary period sites that Stirton, his and Pat's mentor, had explored. We spent several months working with Tedford, Rod Wells and Alan Bartholomai in northern South Australia and southwestern Queensland. We discovered a new fossil field in the Lake Frome

Figure 2.5 Pat in Costa Rica

region, and Pat made two discoveries: the then oldest record of platypuses and the first record of a fossil freshwater dolphin in inland Australia. I was lucky to find a major concentration of fossil waterbird carcasses, for which she and I toiled over making a large plaster jacket that had to make the long trip back to New York to be processed. This was our first experience of 'the Outback', and we were hooked for life. In addition, some of this material was later to influence some of Peter's reconstructions, in particular for *The Fossil Book* (*see* Chapter 4). Once we returned from Australia, both of us not only proceeded with our theses but also became part-time preparators, working on the very material that we had helped collect in Australia, some of which became the source of Peter's reconstructions.

Because Pat was working on a group of gigantic ground birds (the Dromornithidae), primarily collected by Ruben Stirton and his students, after the expedition with Tedford was over, we visited most of the natural history museum collections in Australia so that she could compare her observations based on the Stirton collection with the dromornithid material in those institutions. When we visited the National Museum of Victoria, Alan McEvey, the curator of birds, was particularly helpful and gave her a warm invitation to return and continue her studies of Australian fossil birds if she could find the funds. Later she applied for a Fulbright fellowship and was successful – so we were back in Australia late in 1973.

Figure 2.6 When fieldwork is not exactly fun: getting a field vehicle out of a bog during a 1971 expedition led by Richard Tedford from the American Museum of Natural History in New York. From left to right: Rod Wells (Australian graduate student), Richard Tedford, Tom, and Alan Bartholomai, then director of the Queensland Museum (photo by P. Vickers-Rich).

We both completed our PhD degrees at Columbia University in 1973. Before graduating Pat was offered a job at Texas Tech University and the Fulbright to return to Australia – both on the same day! She decided to do both *and* convinced the management at Texas Tech that if she completed one semester of teaching (and passed her thesis defence in November of that year) and returned to teaching in late 1974, she could take the Fulbright. She did just that. Without a job, and having in the end worked on North American hedgehogs for my PhD dissertation, I accompanied Pat to Australia – basically along for the ride.

One Saturday, after having been in Melbourne for about a month, I bought the *Weekend Australian* newspaper with the intention of reading it in its entirety in order to understand what was going on in the country. Getting finally to the employment section I saw, right in the centre of the page, an advertisement for a curator of palaeontology at the National Museum of Victoria. The advertisement had been limited to Australia. What were the chances of my being in Australia on the day the advertisement appeared, much less of my buying a copy of the *Weekend Australian* and looking in the employment section? Jobs like that were not exactly common. The mind boggles at the improbability. Almost on a dare, I applied for the job and, lo and behold, I got it. This began two years of lengthy separations for Pat and me. She continued her job at Texas Tech, and I worked at the National Museum of Victoria. In 1976, we finally decided that our future lay in Australia. Pat's graduate students in Texas had completed their work, she had taught undergraduate classes for more than two years, and so we moved to Melbourne. With the encouragement of Jim Warren, then the Professor of Zoology, Pat took up a part-time tutorship at Monash University, having given up a full-time, tenure-track position at Texas Tech. Over the next 20 years, she rose to the rank of professor and founded the Monash Science Centre, where she remains today as its director.

In the years between 1974 and 1978, we led numerous field expeditions into Central Australia in search of new fossil mammal and bird sites, with the focus on discovery of sites older than those found by Stirton and his colleagues. Working out from Stirton's and Tedford's sites, we did find new localities and a variety of new fossil species. However, older sites eluded us, though we did gain significant experience in making each expedition cover new ground and narrowing the search pattern. Our first child,

Figure 2.7 Pat and Tom as professional partners, National Museum of Victoria, 1980s.

Leaellyn was born in 1977 and, like her brother Tim, born in 1985, she was out in the field with us by the time she was three months old – what better place to raise a child!

A pivotal moment in our search for the ancestors of Australia's mammals and birds came in 1978 when two young students, John Long and Tim Flannery (both now well known as palaeontologists and authors and Tim known as a campaigner on environmental, social and scientific issues), discovered vertebrate fossils in Early Cretaceous fluviatile sediments near Inverloch in southeastern Victoria. Their discoveries prompted an extensive prospecting and excavation program along the Strzelecki and Otway coasts of Victoria that continues to this day (see Chapter 7).

For most of the last two decades of the 20th century and the first decade of the 21st, the Victorian dinosaurs were the centrepiece of both Pat's and my research. Although Pat's PhD research concerned the fossil birds of Australia, she turned more and more to Mesozoic research. Even though we continued to search for ancestors and ancient relatives of the birds and mammals of Australia, we did not finally have success until 1997. Prior to that, our research on the dinosaurs was central. We worked on what we had. Throughout this time, the National Geographic Society steadfastly supported our fieldwork, and for most of the time the Australian Research Council and Monash University provided substantial funds for the fossil preparation. My position as a curator at Museum Victoria gave me the facilities and the security to concentrate on this research, as did Pat's position at Monash.

Figure 2.8. The Monash Science Centre, at Monash University's Clayton campus in Melbourne. Pat founded the Centre in 1993; it took her nearly 10 years to raise the funds for its construction. This environmentally respectful building provides space for exhibitions and staff, and for the Monash Sustainability Institute, which joined the MSC in 2007. The purpose of the MSC was to connect research scientists with primary and secondary school kids and give some in-depth content to both them and the curriculum in science (photo by Tim Abbott).

Other projects related to our Victorian work took us to Mesozoic sites in southern Africa and Argentina. We also visited many other parts of the world, including Japan, China, New Zealand, Russia, Alaska and Saudi Arabia, where we carried out fieldwork and inspected collections. Our focus, however, remained on the polar faunas of the late Mesozoic of southern Australia.

In the 1990s and early 21st century Pat put a major effort into founding a public science organisation: the Monash Science Centre (MSC), based at Monash University in Melbourne, a place where students, research scientists, parents and teachers could all meet – with an aim to improve primary science curriculum and encourage students of all ages to explore science. She finally secured the $4.5 million to build a facility for both classroom education and exhibitions. Following on from her experience in helping to organise the 'Dinosaurs from China' (see Chapter 5) and later almost single-handedly the 'Great Russian Dinosaurs' (see Chapter 7) projects, she used the funding generated from these and other projects to underwrite the MSC. She finally managed to build and launch the MSC building in 2002, and the University underwrote a visiting scientist position for the inaugural year. Two scientists shared that position: Alexei Rozanov and Mikhail Fedonkin, both from the Paleontological Institute of the Russian Academy of Sciences in Moscow. Many of the casts of fossil vertebrates in the MSC's premier exhibition were the result of this institute's interaction with the MSC and the Queen Victoria Museum and Art Gallery in Launceston, Tasmania (QVMAG), directed by Chris Tassell. Chris was a great facilitator of so many of the programs that Pat organised during the 'Great Russian Dinosaurs' project, and as a result many of the originals of the artworks Peter developed over the years have been donated to the QVMAG, an institution with a true appreciation of both art and science. All three institutions had cooperated significantly, with a great deal of flexibility, in getting the 'Great Russian Dinosaurs' to Australia in the early 1990s, during the period of *perestroika* in the former Soviet Union.

The research interests of Rozanov and Fedonkin were in Precambrian palaeobiology, and during their tenure at Monash they discussed their finds at length with Pat. She was completely fascinated by this field, so new to her, and it was a seminal moment in her life – she left the Phanerozoic (almost) and moved rapidly into a new area – in search of the origins of animals (see Chapter 11). Two years later, she and two of her new colleagues, Mikhail Fedonkin and James Gehling of the South Australian Museum, submitted a five-year grant proposal to study the origin of animals, along with more than 100 of their colleagues. This led to UNESCO's International Correlation Project 493, still ongoing, which resulted not only in significant new international, interdisciplinary research projects, but also in books, conferences, stamps, documentaries and, importantly, funding for a series of palaeo-reconstruction art pieces by Peter (see Chapter 11).

Pat's and my work, over more than 30 years of chasing the scraps of ancient life, chasing our 'dragons' of childhood, and trying to put them into some environmental and evolutionary perspective, has been fundamentally enriched by working closely with Peter Trusler, who is not only a fine artist but also a scientist with an enquiring mind. He can take what is inside our heads, reorganise it in his and bring it to life with pencil and brush. There is so much we would never have discovered had it not been for the meeting, at an exhibition in Melbourne so long ago, of an artist and two scientists.

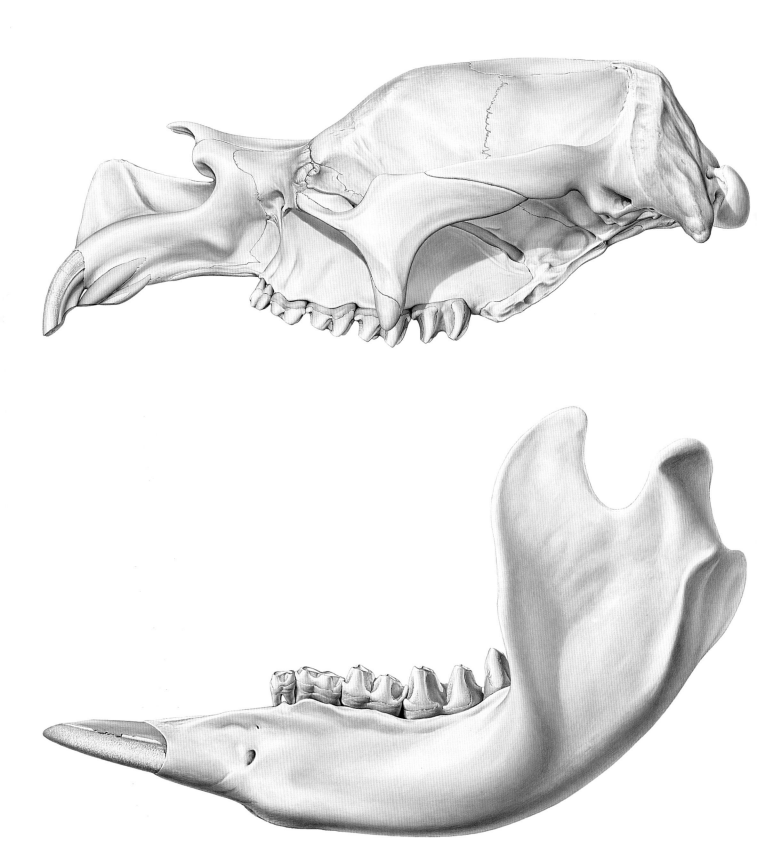

3

The Beginning: Bacchus Marsh *Diprotodon*

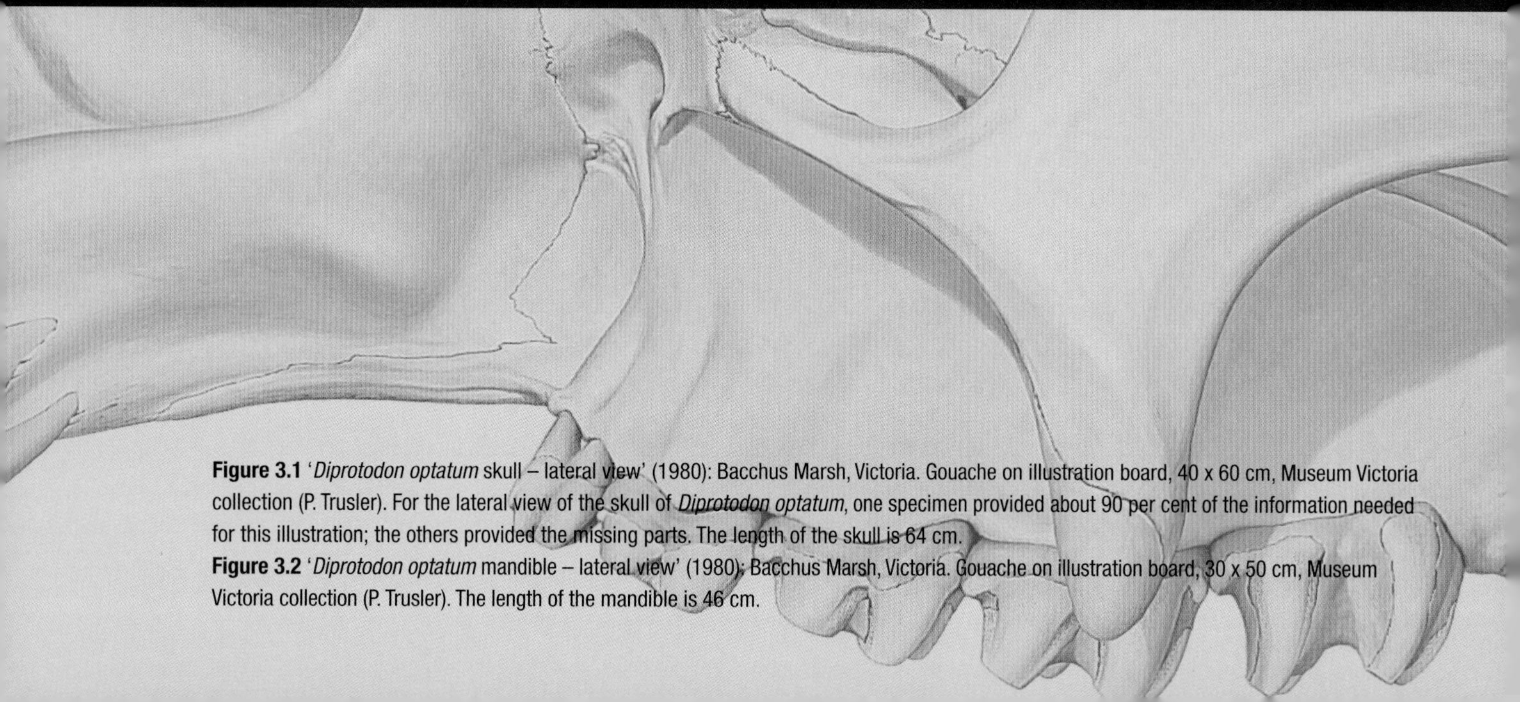

Figure 3.1 '*Diprotodon optatum* skull – lateral view' (1980): Bacchus Marsh, Victoria. Gouache on illustration board, 40 x 60 cm, Museum Victoria collection (P. Trusler). For the lateral view of the skull of *Diprotodon optatum*, one specimen provided about 90 per cent of the information needed for this illustration; the others provided the missing parts. The length of the skull is 64 cm.
Figure 3.2 '*Diprotodon optatum* mandible – lateral view' (1980): Bacchus Marsh, Victoria. Gouache on illustration board, 30 x 50 cm, Museum Victoria collection (P. Trusler). The length of the mandible is 46 cm.

The Scientist: Tom

'Why don't you work on the Bacchus Marsh *Diprotodon?* I am sure that the Australian Research Grants Committee would fund it.' This was what Tom Darragh, acting director of the National Museum of Victoria, said to me in 1978, after I had been working at the Museum for four years. In that time, together with Pat, I had been searching for Australian mammals and birds older than about 20 million years, the oldest found up to that point. But nothing of these older mammals had been discovered, and I wanted something to do scientific research on. I was looking for a project!

Figure 3.3 Sir Richard Owen's 19th century illustration of *Diprotodon*, in which the feet of this giant marsupial are hidden in the vegetation. No one at this point had any idea what the foot structure of this large beast looked like.

Diprotodon was the first fossil mammal described from Australia. This occurred in 1838 when the world-renowned comparative anatomist Sir Richard Owen described some dental fragments from the Wellington Caves in New South Wales. Over the next 40 years, the skeleton of this beast was gradually pieced together – the largest marsupial that ever lived. Skulls were found, but they were always crushed, because their bone structures were incredibly thin, often only one-tenth the thickness of a correspondingly large placental mammal such as a hippopotamus. The Bacchus Marsh *Diprotodon* skulls were not crushed and that was the main reason why research on them was such an attractive proposition.

The specimens Darragh had spoken of had been collected five years earlier, prior to my taking the position as curator of palaeontology at the Museum. The fossils had come to light because of a keen observation made by 12-year-old Kerry Hine while playing in a clay quarry operated by her family at Bacchus Marsh, west of Melbourne.

Figure 3.4 Excavation and plaster jacketing of *Diprotodon* material at Lake Callabonna, South Australia, in 1983 by a team from the National Museum of Victoria, Monash University and the Australian Defence Force (photo by P. Vickers-Rich).

The bone she spotted turned out to be that of *Diprotodon*. She showed her new find to her science teacher, and he in turn notified the Museum. So began a collecting operation extending over several months, carried out primarily by Chris Tassell, then the deputy curator of palaeontology, and Ian Stewart, his departmental technician.

When I took up my job at the Museum, I quickly became aware of this material. However, as my interest lay with older fossils, I did not immediately follow up with a program of preparation and study of these unique specimens or investigate them further. So, for nearly four years, nothing further was done about them.

Following Darragh's prudent advice, I did submit an application to the Australian Research Grants Committee, and they did fund the project, valuing the fact that further work needed to be done. Funding was first used for preparation of the fossils and

Figure 3.5 Bacchus Marsh Quarry area, west of Melbourne (photo by C. Hann).

subsequently for their illustration, and that is where Peter Trusler came to the fore. As the preparation was getting underway, I decided to visit the site where the fossils had been excavated, just to see if there were more specimens to be found and collected. Clearly there were, and so I organised an expedition to do just that. Excavations stretched on into the summer of 1980, when the last dig was organised by Cindy Hann and Greg McNamara, students of Pat at the time. While they were working, yet another project was getting underway, which delayed the research on the Bacchus Marsh *Diprotodon*. Bones had been found in the Mesozoic terrestrial rocks near Inverloch in southeastern Victoria – exciting to me, for a main research interest of mine was Jurassic and Cretaceous mammals. Thus, *Diprotodon* was set aside for a while. When the specimens from Bacchus Marsh were out of the ground and safely in the Museum, they needed to be prepared before any research could be carried out.

Owing to their structural delicacy, and their lack of full permineralisation, they needed to be prepared by experienced technicians. Pat and I had both held down part-time preparators' jobs in the final years of our PhD work at the American Museum of Natural History (AMNH). At the time the *Diprotodon* fossils were to be prepared, I had a full-time job at the Museum, but Pat was only working part time at Monash. With Pat's skill finely honed by her training and experience at the AMNH, she was given the job of chief preparator, which made up her other part-time job.

At that time we lived in Emerald, a rural town quite some distance from downtown Melbourne, where the Museum was located. The commute was at best a 3–4 hour round trip each day. So, with our young family and the responsibilities that go with that, and a limited income, we converted a shed on our property into a preparation laboratory where this work could be carried out with the greatest efficiency of time. The Australian Research Grants Committee (ARGC) underwrote salaries for Pat and our then neighbour, Tim Flannery, who had worked with us collecting fossils since 1974. Pat passed on her skills to Tim, who simply had a natural talent for the job. The time saved by the two technicians not commuting ensured the most efficient use of the ARGC funding.

Once prepared, the fossils needed to be illustrated and thoroughly documented. Because the specimens were fragmentary, the final images needed to assimilate information based on more than one individual. They had to be a composite, not a simple rendering of a single specimen. It was fortunate that all the specimens appeared to be about the same age. In each specimen, the most posterior molar was just coming into full eruption. This stage of eruption suggested that these skulls were

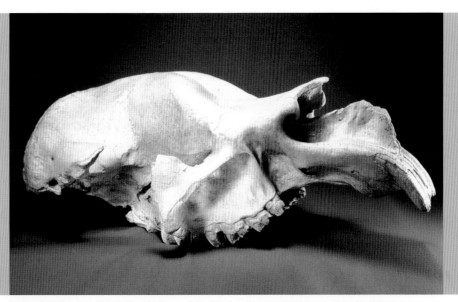

Figure 3.6 Bacchus Marsh *Diprotodon* skull in side view. The length of the skull is about 64 cm (photo by F. Coffa).

from young adult individuals of prime breeding age. In addition, the skulls were all quite gracile (slender), so it seemed likely that they were all of one gender, presumably female. For this reason, combining the information on the morphology of more than one individual into a single, composite illustration seemed a sensible course of action.

Why would a group of 13 individuals, all of approximately the same age, die together as these apparently had? I remembered reading about a study that was carried out on a population of red kangaroos in Western Australia. The observers noted that during a severe drought, the last kangaroos to die were those of prime breeding age, the strongest of the lot. So, based on this modern analogy, I thought it was a plausible scenario for this group of *Diprotodon* specimens. They could well represent an accumulation of individuals that perished in a prolonged drought, well after all the young and old individuals had died. The sediments gave us some further clues about conditions at the time the *Diprotodon*s perished. The sediments in which they were buried were quite coarse, suggesting that the herd was buried by a flash flood event. Either they had died slightly before this event or, in fact, they had been overwhelmed and buried within the sediments carried by this sudden and deadly rush of water.

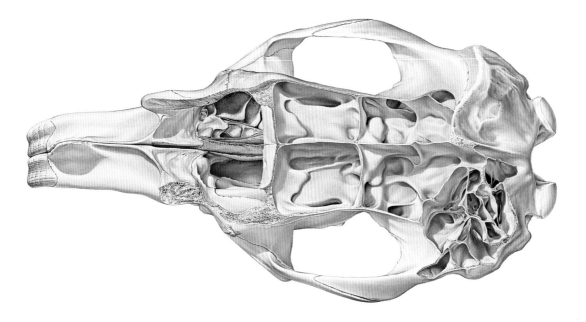

Figure 3.7 '*Diprotodon optatum* skull – dorsal cut-away' (1981): Bacchus Marsh, Victoria. Gouache on illustration board, 40 x 60 cm, Museum Victoria collection (P. Trusler). Skull of *Diprotodon optatum*, looking down on the top with the outer most layer removed. This view was based primarily on a single specimen that had been accidentally cut open by a bulldozer and fortuitously made Tom aware of the intriguing internal structure that just begged for an explanation of its significance. For comparison, similar cut-away views were prepared of the internal skull anatomy of modern marsupials and larger placental mammals (see figure 3.8). The length of the skull is 64 cm.

I realised that a series of composite illustrations was needed. My next challenge was how to find a way to produce such complex images. Melding photographs was not regarded as practical. What I needed was an artist who could produce a camera-quality image by mentally combining those images, taking on board all of the fragmentary specimens and coming up with a realistic, and scientifically accurate, single image. But where was I going to find such an artist?

I learned of an exhibition about to be opened by the Wildlife Art Society of Australasia. If an artist with the desired talents existed at all in Victoria, I thought that this exhibition was a most likely place to find him or her. Pat and I went along for the launch of the exhibition. When we strode into the gallery that evening, where the works of a number of Australian and international artists were on display, we were almost immediately drawn to a bird painting. We stared at it, wondering if the bird was about to fly out of the canvas! We both agreed immediately that our artist had been found. Now, would he agree to work for us? I introduced myself to that artist, Peter Trusler, and asked if he might be interested. In very few minutes, a tentative agreement to work together had been reached. I was ready to move ahead on this, and I think Peter felt the same way. It was a novel project to him – one unlike any he had experienced in the past.

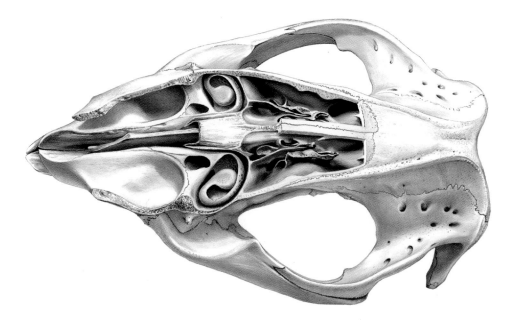

Figure 3.8 '*Vombatus ursinus* skull – oblique cut-away view' (1981). Gouache on illustration board, 25 x 35 cm, Museum Victoria collection (P. Trusler). To better understand the structure of the nasal region of *Diprotodon optatum*, Peter prepared cutaway drawings of the skull of a modern wombat to show the internal structure at the front of the skull. The length of the skull is 17 cm.

In the course of the preparation and subsequent illustration, it became apparent that the skulls of the Bacchus Marsh *Diprotodon* had a rather unusual structure of the internal bones that lay between the brain case and the outer parietals and frontals on the top of the skull. In most cases in mammals, where there are thin-walled laminae within a skull, they are quite convoluted. However, in the Bacchus Marsh *Diprotodon,* there was a flat, plate-like lamina oriented in the sagittal plane (from front to back) of the skull. Perpendicular to this lamina was a second flat, plate-like lamina oriented in the transverse plane (across the skull). The structure of the skull seemed as if it had been designed by a human engineer – placing structures in an orientation where there was optimum use of minimum components (bone) to provide maximum strength. The tantalising question these skulls posed was: why, in this particular animal, had such a novel solution to a problem of providing internal support to the skull developed, quite in contrast to the usual condition present in mammals? At least to me, answering this question was the most intriguing aspect of the study of the anatomy of the *Diprotodon* from Bacchus Marsh.

Before the advent of computers, the way that mammalian skulls were analysed to answer questions of this nature was to consider a very simplified model. The model consisted of plates joined together after the fashion of the actual skull being examined. Then the interactions between the plates in response to forces put on them was calculated. Computer techniques made it feasible to divide the skull model up into a far larger number of plates or elements and calculate their interactions, giving a result based on a smaller degree of approximation. The number of elements for which the interactions could be calculated, and thus the level of refinement, was limited only by the amount of computer time devoted to the project. As that became cheaper, this mode of study – termed finite element analysis, or FEA – became more feasible. In parallel with this, scanning techniques advanced to the point where it was possible to automatically record an image digitally to a very high degree of refinement, rather than having to use crude approximations of the form of a mammalian skull as a combination of simple plates.

This technique of finite element analysis was being developed during the 1960s as a possible way of harnessing the capabilities of computers to deal with problems of this nature. Being a computer programmer at the time, I was aware of this research program. For this reason, when confronted with this interesting problem 15 years later, FEA seemed to be the most appropriate strategy for answering this intriguing question about the skull of *Diprotodon* in the most reliable way possible. So, after a patient wait of three decades for technology to advance, it is now possible to complete the study of the Bacchus Marsh *Diprotodon* in the manner originally conceived.

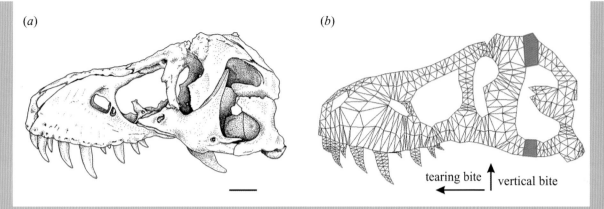

Figure 3.9 When a force acts upon a structure, if the structure can sustain the load, the external force is counterbalanced by the structure itself. Internally within the structure, the pattern of counterbalancing forces is dependent upon both the shape of the structure and the nature of the material or materials out of which it is made. A method to determine the direction and magnitude of those internal forces is to consider the structure divided up into a large number of elements and to then calculate the resultant forces acting upon each element by its interaction with the adjacent elements. This technique is called Finite Element Analysis. (a) shows a conventional image of a skull of *Tyrannosaurus rex*. (b) shows a network superimposed on such a skull. Each triangular area is an element. By studying the direction and magnitude of the internal forces acting on each of these elements, it is possible to understand various aspects of the structure, be it a bridge or a skull. From fig. 1 in Rayfield (2004) used with permission. Scale bar 10 cm.

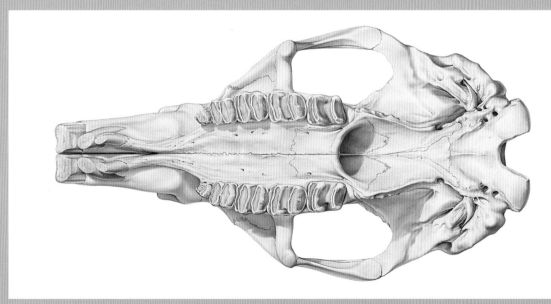

Figure 3.10 '*Diprotodon optatum* skull – ventral view' (1980): Bacchus Marsh, Victoria. Gouache on illustration board, 40 x 60 cm, Museum Victoria collection (P. Trusler). The length of the skull is 64 cm.

The Artist: Peter

An evening soiree was the occasion that saw my introduction to Tom Rich. He quietly moved towards me from the midst of a crowd of gallery goers at the 1978 annual exhibition of the Wildlife Art Society of Australasia in Melbourne.

'Hi! You're Peter Trusler? ... I'm Tom Rich!' he exclaimed, in his clipped American manner.

I guess Tom was readily able to identify an artist – this awkward-looking, long-haired 1970s fashion victim – from a more elegantly presented crowd. He had correctly pinned me to the particular painting that had taken his eye. Unbeknown to Tom, I had recognised him in advance, for I had known of both Pat and Tom for some time.

I had been aware of their work and interests through a couple of sources. Firstly, as a final year undergraduate student at Monash University, I had attended occasional seminars that they had presented on palaeontology. Secondly, I had also been an occasional volunteer on fossil digs. Thus, we each had an association with Professor Jim Warren, the head of the Department of Zoology of Monash University at the time, as well as some of his technicians and graduate students – one being Ken Simpson, a keen birder, and another Gordon Sanson, who worked on the functional morphology of kangaroos.

The Curator of Ornithology at the Melbourne Museum, Alan McEvey, had a passion for art, and this provided another connection. Alan had been quietly supportive of my painting interests since I was about 14 years of age. I had been talked into sending him a little watercolour sketch I had done of a raptor that I couldn't identify, observed while on a family holiday at Port Campbell National Park on the spectacular Victorian southwest coast. My father and I had asked the local park ranger to identify some birds for us, based on my sketch. The ranger, in turn, called over the director of the Parks Service, who just happened to be in the local office at the time. This was Dr. Len Smith, himself a well-respected ornithologist, who immediately identified the bird and added: 'It would be very wise if you sent the drawing to Mr Alan McEvey at the National Museum of Victoria to confirm the identification. He would be very interested to see what you have done!' I recall being somewhat taken aback by this 'official' interest and dutifully posted the scrap of paper. I had been witness to some marvellous aerobatics in a family of swamp harriers and did not realise that the 'fighting' I had observed between a number of distinctly differently plumaged birds was in fact the soaring interaction between a pair of adult harriers and their three darkly-coloured offspring from the previous season. Mum and Dad were evicting the teenagers!

Alan McEvey would have been among the crowd that evening too, for he had been a primary instigator of the Wildlife Art Society of Australasia's formation and its subsequent development. It was he who had involved me with this group from the outset. He had also been working professionally with Pat on fossil birds.

From the brief discussion that Tom and I had about my painting, I gleaned that he had been taken, not so much by the detail of my technique or his assessment of matters of biological accuracy, but more particularly by the multi-dimensional illusion of the realism with which I had been experimenting.

This style had stemmed from a number of sources, but primarily from the tuition I received from a local artist in Ballarat, who in turn was a pupil of Max Meldrum, a cantankerous Scot obsessed with the science of appearances and a resolute anti-modernist. His charismatic career had spawned an industry of traditional realism in the conservative environs of Melbourne from the 1930s to the early 60s. It was also an aspect of academic art practice that had long fallen from favour in the secondary and tertiary education curricula. However, thanks to the patience of one of his many pupils, my teacher, Jessie Merritt, I had absorbed some very fundamental training in representational practice and perception at a very early age. Art tuition for children was her forte, and I was a willing sponge.

Tom had come on a quest for a particular kind of illustrator, one with skills that could be brought to bear on a specific project … Enter the *Diprotodon*! Would I be interested?

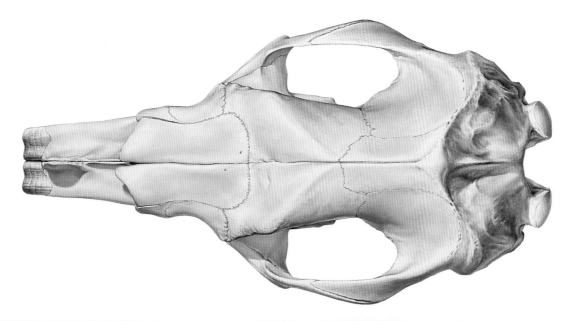

Figure 3.11 '*Diprotodon optatum* skull – dorsal view' (1980): Bacchus Marsh, Victoria. Gouache on illustration board, 40 x 60 cm, Museum Victoria collection (P. Trusler). The length of the skull is 64 cm.

Tom wasted little time on the effusive superlatives that one is inclined to receive at the wine and cheese-infused social events that are exhibition openings. 'I like what you do,' he concluded. He enquired further of my background and offered some about his, elaborating mostly about the issues that confronted him with a new fossil discovery at Bacchus Marsh, some 40 km (25 miles) west of where we were standing in the corner of the upstairs gallery of the old Vic Arts building in Albert Street. I can clearly remember that when I mentioned being an undergraduate student in zoology at Monash, his eyes swelled to fill the rims of his spectacles, and a broad smile rimmed his American-style beard as he exclaimed, 'Oh!' The incongruity of this remark later made me realise that for him, educating this young artist was not necessarily going to have to start from scratch. Maybe the possibility of my possessing skills in both disciplines was something he had not immediately expected? What luck! The doors to the requisite lines of communication were open, and so a fascinating journey was about to begin. I really had little idea of what I was getting into, but I was intuitively enthusiastic about the prospect of being rigorously challenged and the fact that someone thought I might be capable of passing such an esoteric test.

Art and Illustration

A few introductory remarks about the possible distinction between these two terms might be appropriate here, because it is pertinent to my involvement in matters scientific. This indeed was a favourite point of discussion between Alan McEvey and a number of members of the fledgling Wildlife Art Society of Australasia. Alan used to continually provoke me on such issues whenever we stood before a piece of work, and I learned much from him, so much so that we often became entangled in intense discussions long after fellow members had departed meetings. His own arts writings had centred on the works of John Gould and the genre of 19th century natural history illustrators, who continue to have a strong influence over the field to this day.

The setting for my first meeting with Tom was, therefore, one centred within a society whose members were interested in natural history, first and foremost. The majority of its members were practising artists – some professional, many recreational – and, aside from a number of educators, most survived largely by the production of decorative paintings for private sale, commercial gallery exhibitions or illustration commissions. A few years before this, I had taken a break before deciding whether to pursue postgraduate studies, and had commenced freelancing as a painter. This was financially fatal, of course: I had done a little of anything-and-everything as artists normally do to survive, but was spending much of my time painting with no immediate purpose in mind.

Throughout this formative period and my connection with this Society, there was a much-needed impetus to move from work that was fundamentally representational or figurative towards work that was more aesthetically inspired and more conceptual. Fundamental issues between tradition and invention were fought and lost, but there was a genuine craving among many of the Society's members that the wellspring of change needed to come via an evolving culture, experimentation that stemmed from an expression of the human spirit. Did I understand what this might mean? The subject fascinated me. Art fascinated me, but did such a distinction imply that realistic art in any guise was a contradiction in terms? Was wildlife art, as it was commonly known, simply an illustrative genre? Perhaps it was a quiet backwater in which reclusive animal lovers could admire or create images of the objects of their fascination? Alan didn't always think so. Nevertheless, I still feel that when art practice is purely subject-based, it tends to be purely illustrative.

An analogy might make this clear. How might one transcend from a painting of a B-class locomotive engine, something that a railway enthusiast might admire, to a work of the calibre of English Romantic painter Joseph Mallord William Turner's *Rain, Steam and Speed* or one of Claude Monet's paintings of the Paris railway station? It is not simply the passage of time, some perceived earnestness or dexterity of the loose brushwork or a consensus from generations of art commentators and historians. Values by any measure can be fickle. Why might that consensus become so profound? More important to me at that time was: why was I attracted in the opposite direction? This was certainly so 'un-cool' in the socially accepted artistic parlance of the time. It also occurred to me that, irrespective of the accessibility of science in a popular sense, or the depth of its theoretical rigour, there was not necessarily going to be a direct correlation with the quality of the science or the quality of the imagery used to communicate it.

I am not entirely certain why I was attracted to this sort of art rendering. I have developed personal psychological assessments over the years of contemplating such issues, but essentially I viewed myself as requiring far more technical development and perceptual training. I lacked drawing practice. I needed more study of art history and anthropology. I wished to become more competent at design. I needed more study of biology, of anatomy and certainly more knowledge of the way in which natural systems work, both physically and aesthetically. I actually saw this as a counterweight to the progression of the intensely personal painting that I had begun to pursue. I really wanted to see if I could truly paint and was making some headway, only to reach a new level and become aware that some other aptitude was wanting. There were so few avenues available at that time to follow such diverse interests without having to be supported as a student. My painting barely supported me full-time, and certainly could not support me while studying part-time. I did not have the requisite background to compete for arts grants. So, here was someone willing to pay me to satisfy some of these conflicting cravings through a very unusual project. Tom had my attention. I was intrigued. Probably an opportunity not to be missed, I somewhat naively thought, and possibly a means to a variety of ends.

When I arrived at the Museum some time after our initial discussion, my eyes widened with incredulity! In drawer after drawer of antique blackwood cabinets, laid out on rolls of plastic across countless tables and in neatly sorted piles of newspaper and plaster jackets spread over the beautiful parquetry gallery flooring, were hundreds of lumps of whitish chalk. These white rocks were scarcely discernible from the huge fragments of hardly fossilised bones. As Tom wandered about expressing a sense of preciousness about this mess (and he had obviously been struggling to contain it, for there was musty-smelling white dust everywhere), I gradually began to see the forms that were emerging from the chalky clay. He led me over to a side wall of temporary shelving, and there they were: 13 massive skulls in various states of completeness

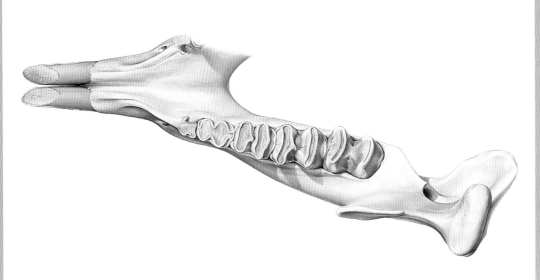

Figure 3.12 '*Diprotodon optatum* mandible – occlusal view' (1981): Bacchus Marsh, Victoria. Gouache on illustration board, 30 x 50 cm, Museum Victoria collection (P. Trusler). The length of the mandible is 46 cm.

and at advanced stages of preparation. This was the raw material on which I had been invited to work.

Tom continued his monologue with increasing enthusiasm, pointing out features here and there and making an incomprehensible range of comparisons between one skull and the next. Unbelievably, the bone was as crumbly as the clay that encased it. What was more, it was virtually the same pallid, cool grey, and I was struggling to keep up. Tom spoke of the techniques and difficulties of extracting the specimens and the methods that he, his staff and volunteers were using to prepare them. He candidly described what was working well and where they were running into problems. They were developing ways of improving the extraction of the specimens, and Tom was expressing grave concerns for their long-term preservation. Clearly, my work was going to be a major clause of his insurance scheme for this collection.

I was immediately taken by the exposed rows of glistening white teeth, bearing their characteristically textured enamel. The massive, curved front incisors lacked enamel along the length of their lateral faces. The smaller incisors were merely dentine pegs. On closer inspection, the cleaned bone surfaces revealed the finest detail – beautiful meandering suture lines, tiny pores and soft sculptural curving forms bearing deep foramina. A whitish anatomical landscape was emerging before my eyes, and it seemed as fresh as the day it was buried.

At one point during this initial survey, Tom gestured with a sweeping hand, 'The bulldozer found this one. It's kinda neat.' I was stunned. There was an entire skull, from which the dorsal surface had been neatly sheared. It was a perfect section that revealed the entire internal anatomy with the same fidelity as the external surfaces, a labyrinth of sinuses surrounded by thin walls of bone. All was unbelievably fragile and yet, for the most part, un-collapsed. The anatomy seemed biologically implausible to me. It was equally amazing that the preparators had so patiently and dexterously removed the entombing clay. I did not know until later that the clever fingers that had prepared this material belonged to Tim Flannery and Pat. That clay they had slowly and meticulously removed had ensured the delicate skulls' survival until now. Making moulds of these specimens was going to be a risky business and not an option that Tom was willing to take until the art documentation was complete, and maybe not even then.

Tom and I discussed the project. The skulls represented upwards of 13 individuals, possibly from a single population, whose lives were abruptly terminated in their prime. Their ages, as indicated by their teeth, ranged from the youngest individual, whose last molar had not fully erupted from its jaw, to the eldest individual, which displayed only moderate wear facets on the equivalent tooth. The majority of the herd had just matured, and these individuals all represented a somewhat smaller, more gracile form than the massively bulbous-nosed specimens I had seen elsewhere. Tom inferred from that fact that they might have been females, and their age cohort was suspiciously like that recorded for modern kangaroos at the end of a prolonged drought. Diseased and aged individuals tend to suffer the effects of drought first and are quickly culled from the population. The population simultaneously reduces breeding activity. This increasingly biases the age ratio towards the fittest young adults, but after a time the conditions exact a toll on these individuals too. Palaeontologists are all too familiar with collections represented primarily by the young and the old, and so, as I was to become aware, they get excited about prime specimens. They see things with different eyes.

These were important observations, but regardless of the reasons for the accumulation of these deposits of disarticulated bones, the narrow age range provided a unique opportunity to compare all individuals with greater assurance. My task need not be strictly a standard recording or an illustrative one, because there was likely to be observations that applied to them all. This enthused me further.

In my earlier efforts to hone my pictorial skills, I had been humbled by the generations of engravers and lithographers who devoted their lives to painstakingly recording biological material. From my childhood birding interests, the popularity of artists such as J. J. Audubon and Joseph Wolf was inescapable, but anyone from Albrecht Dürer to George Stubbs held my attention. During the 1970s, the increasingly republished biological illustrations from texts such as *Reports on the Scientific Results of HMS Challenger* (1887) directed me to works by Ernst Haeckel and so many other specialist illustrators. On many occasions I had pored over the plates from Sir Richard Owen's compendium *Researches on the Fossil Remains of the Extinct Mammals of Australia with a Notice of the Extinct Marsupials of England*, published by J. Erxleben in London during 1877. Therein were exquisite renderings of *Diprotodon* fragments. I could strive for that level of excellence, but Tom was not asking me to walk down that path. He clearly wanted much more!

With respect to art in general, Pablo Picasso is reported to have exclaimed that photography had thankfully released us from those shackles of realistic representation.

So true, but for all major advances there is a frame of reference, or a point of relativity, beyond which this may not hold. Regrettably, something is often lost when another thing is gained. I often bear this cliché in mind. In circumstances where I have the opportunity to compare a well presented, old lithographic drawing of a specimen with an equivalent photograph, I have found that, contrary to popular opinion, the drawing can be superior. There are two reasons for this.

The first is that a drawing is a complex synthesis of information, which embodies a hierarchy of decisions. It contains a system of weighted emphases that can filter out extraneous or irrelevant information. Drawings deal with surfaces, form and content and matters of understanding, for they are time intensive. Drawings are expressions that embody research and development. They are never bland, factual presentations, no matter how simple or realistic they may appear. The specimen's image has been considered, and not shot! One can, therefore, read the mind of the individual who rendered it. This is not to say that an astute photographer cannot effect many of the same processes with skilful camera technique. After all, speed and cost are paramount. Stereo image systems bring other advantages, and digital systems can now expand imagery potential enormously in all respects.

The intrinsic difference for me, as a practitioner, is also one of cognition. Once I draw something and/or take notes about it, the process paves the pathways to memory recall and builds knowledge. If I simply photograph it (and I'm doing this constantly!), I achieve great efficiencies of time and amass enormous detail, but I tend to remember little about the subject. It becomes harder for me to grasp the essence of the whole item or understand it. There is a personal issue of practice going on here too. I have reached a stage where the technical aspects of drawing are second nature to me. I no longer have to think too much about technique, and it has become a natural method of seeing for me. Conversely, with photography, I need to concentrate on the technical processes at hand, and I devote less time to actually looking at the subject. Thankfully, I have slowly improved in this respect, and the additional time I now spend with digital systems has brought me a little closer to my comfort zone.

On the other hand, one of the main issues for the scientist concerns the accuracy of the accompanying artwork. Human skills can be riddled with error. The hierarchy of well-contemplated decisions can be built on a faulty fact or assumption.

Generally speaking, photography is still viewed as factual. It has had such a strong claim to make in this regard that the advent of easily manipulated digital systems is not likely to shake this general conviction for some time to come. Artwork is seen as sloppy. At best, it is a powerful emotional statement, and at worst full of errors and distortions generated by incompetent skills and perceptions or just lapses of concentration. Artists have for centuries relied on all manner of optical and mechanical devices to minimise these problems where they are an impediment to the objective. Does this clearly contradict my argument here? Yes, it can. But there is another trend that runs counter current.

The second reason that drawings can be superior is historical. The increasing reliance on sophisticated photography has so dramatically reduced the need for drawn recorded imagery that the inherent graphic skills involved have diminished too. Draftsmanship has waned. In a broader artistic context, this has not only dramatically

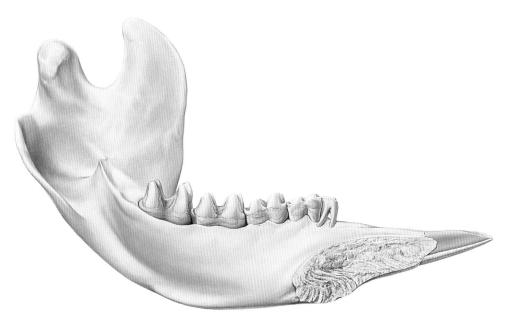

Figure 3.13 '*Diprotodon optatum* mandible – medial view' (1980): Bacchus Marsh, Victoria. Gouache on illustration board, 30 x 50 cm, Museum Victoria collection (P. Trusler). The length of the jaw is 46 cm.

changed drawing stylistically, but also shifted its intent in new directions, as Picasso had predicted. For the purposes of the present discussion, however, it has resulted in a lowering of the average, such that contemporary realistic illustrative drawing can seldom match the powers of photography and reproduction. So much drawing practice is actually now based exclusively on photographic imagery. For the artist at the end of over a century and a half of this interaction, and as a result of all manner of short-term expediencies, the potential for an intellectual contribution through traditional drawing practice has been 'dumbed down'. Just slaves to the instant image? If only I could ask of Picasso: 'Have matters really changed?'

At the time of my introduction to *Diprotodon,* I was somewhere in the midst of all these issues.

The pale grey forms of the *Diprotodon* bones spread before me, naturally lit from the vaulted skylights of this former Victorian exhibition gallery, called to mind a collection of broken classical marble sculptures. Was I going to have the wherewithal to establish their scientific and aesthetic significance? Was I going to be able to surmount the steep learning curve required to complete the task?

Tom and I discussed the value of representing the total form of these skulls as well as the myriad anatomical details. He wanted me to be his eyes, and he wished to escape the necessity for countless pages of complex verbal description. We agreed that there were good grounds for producing a visual summary of all of these specimens, a succinct review of all salient features inside and out, leaving only the variations or exceptions to be accounted for in text. It was to pave the way for matters of functional and evolutionary analysis. We discussed matters of measurement and statistics, and my mind began to melt a little. I could certainly attempt an idealised representation, but I was unsure of the rigour required for the fidelity that he would insist on.

At the conclusion of my existing illustrating commission for *Birds of Australian Gardens*, I made a start.

The Work

The measurement process, as tedious as it was to begin with, was soon a matter of routine. By basing our 'ideal' specimen's age on the degree of tooth wear, where the last molar had worn down by several millimetres when compared to a newly erupted tooth, we assessed the variation and could average measures to this condition. I could construct a series of orthographic projections reconstructing a skull perpendicular to either the sagittal or dorsal plane and thereby eliminate parallax issues. This would be of value and create a series of accurate interrelated images, all of which could be directly measured for most comparative purposes.

However, such a series of 'maps' of the topography would not adequately address the problem of communicating the third dimension. This was looking disappointing, and I started talking with Tom about 3-D imaging techniques. My interest in academic illustration had often taken me to the Biomedical Library at Monash University when I was a student. There, I had whiled away precious study time looking at British and American journals on surgical procedures and a small number of works on medical illustration – that's where the real money was in bio-illustration. To my delight new techniques were being employed to carry out standard illustrating tasks, and I had begun to incorporate aspects of these in my general tonal drawing practice. In a few instances, I chanced on references to 3-D line illustrations as instructional aids for microsurgical procedures, and quickly gleaned enough about the technique to try it out myself. It worked relatively easily. It became increasingly difficult, however, to manage with more complex curving morphologies. The subtleties required to maintain a consistent depiction of topographic relief for the morphology being illustrated became more demanding. The illusion was easy to exaggerate, but harder to faithfully adjust to a one-to-one ratio reality. Furthermore, the dimensions of the specimens on which I was working were large overall; the larger the drawing, the more difficult I found the changes in parallax. The lines were then not meshing well, and the prospect of creating tangible surfaces from two disparate images was too difficult. The images for this technique needed to be relatively small and simplified. I would lose the detail that I sought, and the ability to provide measurable work. The workload involved in continually testing the production of two images for one resultant illusion was insane. All in all, I could not provide a significant advantage over stereo photography.

I returned to my standard tonal drawing theory to determine how best to re-establish a realistic illusion of depth in my measured drawings. My facility with gouache and watercolour painting used for the *Birds of Australian Gardens* illustrations had come a long way. I realised that gouache was a medium ideally suited to capturing

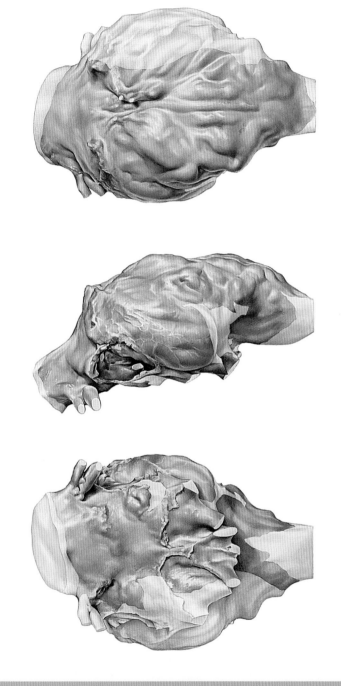

Figure 3.14 '*Diprotodon optatum* – brain case endocast in three standard views' (1981): Bacchus Marsh, Victoria. Gouache on illustration board, 50 x 30 cm, top, side, bottom. The length of the brain is 11 cm (P. Trusler, Museum Victoria collection).

the chalky texture of these skulls. It was fragile to the skin contact inherent with its use, but it dried quickly, could be easily corrected and entirely reworked if necessary. It dried slightly darker than when wet and there were challenges to be overcome when attempting to continuously blend tonal and colour gradations. But my experience with rendering the realism of bird images enabled me to cope. The resulting flat matte surface was perfect for reproductive purposes. Tom needed only monochrome works. This was it. If I standardised the intensity of a hypothetical soft, parallel light source from a consistent angle of incidence across the entire specimen, I could adjust tonal values, highlight and shadow the dimensions to beautifully gain the required depth illusions on the flat surface of the paper.

The internal anatomy had to be approached differently. Aspects of the auditory (hearing) system, for example, needed to be revealed at various depths, and these and other features could only be seen well from non-standard views. The number of possible drawings was increasing, and so in order to maximise efficiencies, Tom and I settled on a few oblique perspective views – once he became confident of the fidelity of my visual 'illusions' of the physical reality. And so I produced a series of cut-away drawings, inspired as much by the stunning engineering drawings that were popular in specialist motoring magazines as by those from any anatomical treatise or architectural drawing. These drawings were presented in the same tonal style as the other measured works.

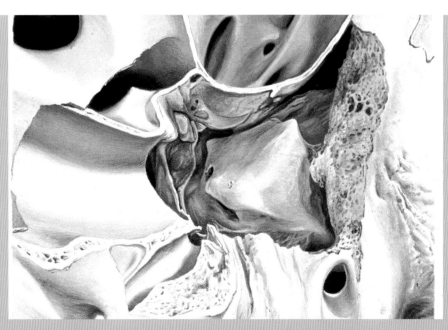

Figure 3.15 '*Diprotodon optatum* auditory region – oblique view' (1981): Bacchus Marsh, Victoria. Gouache on illustration board, 20 x 30 cm, Museum Victoria collection (P. Trusler). Close-up view of the ear region of *Diprotodon optatum* to show the internal anatomy. This area of the skull is one of the most informative for establishing the familial relationships among mammals. The width of the area depicted is 64 mm.

Near the end of the project Tom had an internal silicon cast taken from one of the most fragmentary specimens, which revealed the morphology of the brain cavity without harm to the specimen. I prepared standard views of this cast as orthographic projections, in the same manner as for the external skull morphology. We then started dissecting wombat heads and skulls for comparative purposes, and I produced a matching series of images to demonstrate the homologies in nasal sinus anatomy. We were now starting to think of function and evolutionary pressures that would have led to the anatomy we were observing.

I felt an enormous sense of achievement towards the end of this project, because I had conquered some fundamental issues. I had confronted the gaps between single point perspective, parallel projection and binocular depth perception. I was content that I had produced a series of illustrative works that were as much about a state of knowledge as they were about a collection of anatomical facts or a series of measurements. What is more, I felt that I had dealt with this suite of issues in a seamless way and with an aesthetic sensibility. The challenge to my skills was formidable, but all the shortcomings by virtue of my craft were not diminishing the marvellous collection I was illustrating. And those enigmatic gaps that exist between the physical world and the human perception of it were as few or as small as I could make them. Had I, as I hoped, succeeded in weaving a fiction that represented the truth?

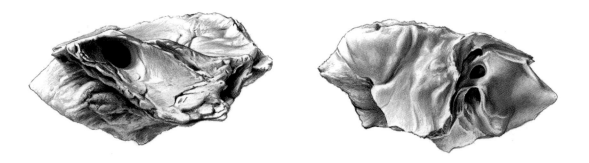

Figure 3.16 '*Diprotodon optatum* petrosal – lateral and medial views' (1981): Bacchus Marsh, Victoria. Gouache on illustration board, 20 x 30 cm, Museum Victoria collection (P. Trusler). Detailed view of the petrosal bone of *Diprotodon optatum*, which housed the sensory tissue for both hearing and balance. The horizontal length is 43 mm.

4
The Art of *The Fossil Book*

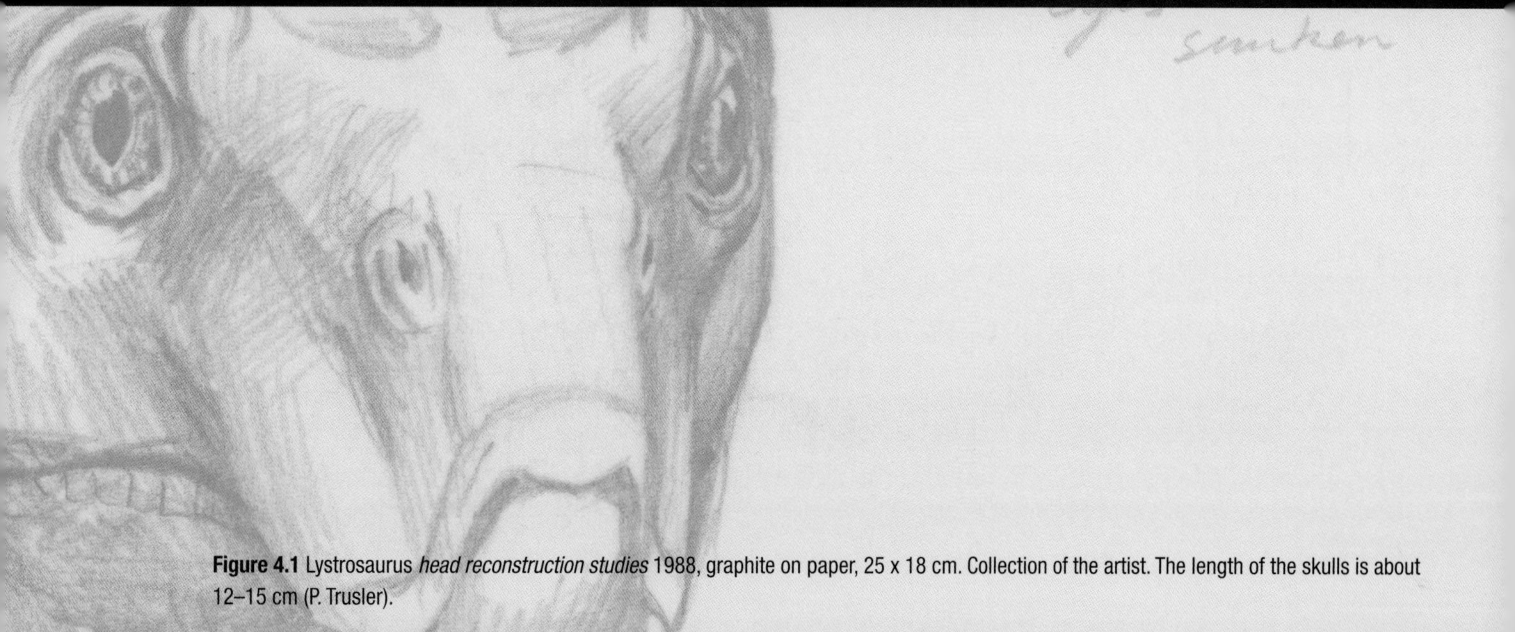

Figure 4.1 Lystrosaurus *head reconstruction studies* 1988, graphite on paper, 25 x 18 cm. Collection of the artist. The length of the skulls is about 12–15 cm (P. Trusler).

The Scientist: Pat

Imagine yourself crouching behind the massive trunk of an ancient pine tree in a cold polar forest of Alaska or Australia – watching a herd of small dinosaurs grazing in the moonlight. Or think of what it might be like to lazily paddle in a warm sea in central Kansas and startle a Nautilus-*like ammonite as it floats near the surface. It jets a cloud of black ink and darts away. Fossils of such animals allow us to travel back into the beginning of life on Earth, more than three and a half billion years ago, when early singled-celled organisms lived.*

The Fossil Book, 1989

I remember writing those words late at night, sitting at a makeshift desk with my little daughter sound asleep in bed on the other side of my desk – she wanted to doze off next to 'Mummy'. I also clearly remember how such intimacy could take place, for had it been 10 years earlier I would have been tapping away on a noisy old 1932 LC Smith Corona mechanical typewriter! Instead, Mildred Fenton, a co-author of *The Fossil Book* I was updating, had generously purchased one of the early personal computers – an Osborne, something of a portable brick, but it was almost noiseless, and certainly conducive to lulling my child to sleep.

The second edition of *The Fossil Book* had its beginning in China, when I travelled to this still Maoist country as a representative of the Australian scientific community. My first impression of Beijing was a mass of bicycles. This was 1979 and hardly a car was to be seen! For a number of years after the Cultural Revolution in the 1960s our Chinese colleagues had been ghosts – we had not heard from them, and we had little, if any, idea of what they were doing and even if they were still alive. Then, quite by surprise, in November 1973, when I was finishing my PhD at Columbia University and the American

Museum of Natural History in New York, I received a postcard from Prof. Minchen Zhou (or Chow), then head of the Institute of Vertebrate Paleontology and Paleoanthropology, Academica Sinica. He noted purposefully that contact between East and West was to be re-established, and the palaeontological community at IVPP was alive and working again.

I finished my PhD and was able after a few years to secure a part-time post at Monash University in Melbourne, Australia. In 1979, as a young academic, I wrote to the Australian Academy of Sciences in Canberra and cheekily asked if I could go as a representative of the government (although I held only a permanent resident visa) to assess the state of palaeontology in China, and to make contact with Minchen Zhou, the head of a major scientific organisation in the Peoples' Republic of China. The answer came rapidly, and was positive. Not only was it positive, but I was asked where I wanted to go. That was easily answered, as I had quite specific interests in several classic fossil sites of Cenozoic and Mesozoic age in China. I was subsequently able to visit every place on my list, except for one – Nei Mongol (Inner Mongolia). I had failed to take note of the season: my visit was to be in December of 1979, and Mongolia would be under metres of snow! No matter, there were plenty of other very interesting places, many of which had not been visited by Western researchers for decades – more than a century in some cases.

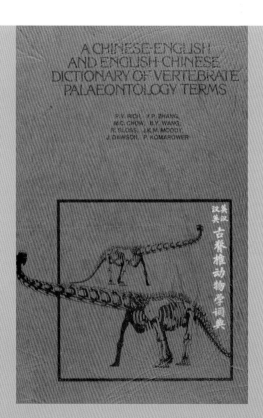

Figure 4.2 *A Chinese–English and English–Chinese Dictionary of Vertebrate Palaeontology Terms*, produced in the early 1980s as a result of Pat's visit to the Institute of Vertebrate Paleontology and Paleoanthropology in Beijing.

The result of this trip was the establishment of strong personal and scientific connections, in particular with researchers at the Institute of Vertebrate Paleontology and Paleoanthropology. Many became dear friends and subsequently visited Australia and lived with my family and me over months to carry out many joint projects. These ranged from the production of two Chinese–English/English–Chinese palaeontological dictionaries to major exhibitions (*Dinosaurs from China*, 1982–85) that travelled to Australia and, of course, a number of research papers – significantly, one journal article by Minchen and Tom on the unusual jawbone of a Mesozoic mammal, *Shuotherium* (*see* Chapter 12). That small jaw introduced to the palaeontological community a whole new group of mammals, so completely different in molar structure to all other members of this group, and perhaps was the discovery of greatest importance that resulted from my first trip to China.

One of the other outcomes of this Chinese–Australian connection was *The Fossil Book*. Minchen, not only the director of IVPP but also a senior and well-respected research scientist, became a collaborator with both Tom and me, as well as a close friend. After lengthy conversations over several months, when we all got to know so much about each other's research and educational philosophies, Minchen said to me: 'You need to update *The Fossil Book*.' This was an interesting prospect, for this was the very book I had grown up reading, word by word, and inspecting, picture by picture.

Figure 4.3 Carroll Lane Fenton (left) and Mildred Adams Fenton at work in the field as young geologists; together they published the first edition of *The Fossil Book* in 1958 (photo courtesy of M. Fenton).

The Fossil Book had been written by a very enthusiastic and professional pair – Mildred Adams Fenton and her husband Carroll Lane Fenton – and had been a bestseller over decades with fossil enthusiasts. It was also used by teachers at many levels, and I could remember using it as a reliable reference when I was studying for my undergraduate exams. It had hundreds of beautiful illustrations of fossils that Carroll Lane Fenton, a PhD himself, had studiously drawn, and accompanying photographs that Mildred had purposefully and masterfully taken, all of the quality one would expect in research publications.

Minchen then said: 'You know, I lived with the Fentons in the US while I was finishing my doctorate before returning to China. You need to get together with her [Mildred, for Carroll had died decades earlier] and update the book.'

The book had concentrated mostly on North American examples to tell its fossil stories. Tom and I set to work with Mildred, who was then in her late 80s. We nearly doubled the length of the book by bringing it up to date and expanding it to include the world fossil record. Maintaining the writing style of the Fentons was a challenge, for Tom and I wrote in an entirely different manner. Still, we took this on with relish, and Mildred seemed pleased.

Mildred was no passive bystander. She travelled to Australia, staying in our home and taking great interest in our daughter Leaellyn, not yet at school. Mildred carried out further photographic work in Western Australia at Shark Bay, visiting structures (stromatolites) laid down by some of the living descendants of Earth's oldest life. She carefully read and edited anything that Tom or I wrote. She became an inspiration to me for the way I would like to spend the last years of my own life.

It was during one of her early visits to Australia that she and I decided we needed some additional reconstruction work; Peter Trusler was the name that immediately came to mind, for his work had impressed Mildred immensely. An academic herself, and with a mind attuned to accuracy, she was in complete agreement with me. We then had to convince our publishers at Doubleday, which was funding this venture, that Peter must be paid a reasonable amount for his time. At first they questioned this, but not after they had had a chance to examine his previous work. In fact, the management at Doubleday were astute enough to realise how important accuracy of scientific detail was, even in a publication that they looked upon as 'popular'. Not all publishers are so enlightened, it seems these days.

I immediately contacted Peter and asked if he would be interested in rendering several illustrations for this new edition – they were to be in black and white, not only to keep the cost of production to a minimum, but also to maintain the style of the original edition of *The Fossil Book*.

Mildred and Carroll had both been trained as palaeobiologists who dealt with invertebrates, and their illustrations dealt well with this realm of life. I had worked primarily on vertebrates, and with this in mind Mildred and I decided that vertebrates should be featured in the new art. Our research colleagues had much to offer, based on our joint research on fossil reptiles, birds and mammals from around the world.

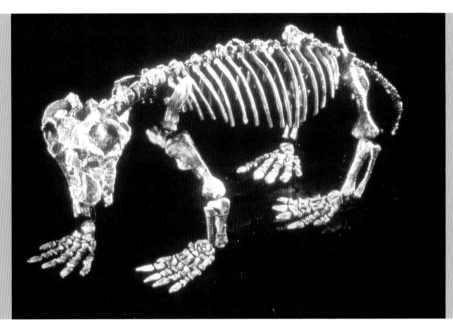

Figure 4.4 Skeleton of the mammal-like reptile *Lystrosaurus*. This was a form widespread across the supercontinent of Gondwana and one of the links used in the early 20th century by Alfred Wegener to connect all these lands in his concept of continental drift. Later it was found that this form actually roamed across an even larger megacontinent of Pangaea, this specimen being from Russia. The length of the skeleton is about 90 cm (photo by F. Coffa).

The work that Tom and Peter had been pursuing for the past few years, on such topics as *Diprotodon* and the Chinese *Shuotherium*, provided the rigour that Peter needed to proceed, and we took full advantage of this backlog of research data. Thus, the background information collection time was shortened, but we also knew that these illustrations would serve multiple uses in the future. The funding that was to be provided by the publisher saved time for me in grant writing to find additional support for Peter's work. The royalties to be paid on the book could be further used to support art and research – such forward thinking was needed in a time of shrinking resources for the communication of science, even in the 1980s.

Peter produced seven illustrations for *The Fossil Book*, beginning with a reconstruction of the very successful mammal-like reptiles and their reptilian prey from a time some 250 million years ago. This tableau included the famous *Lystrosaurus*, a fossil saurian that had allowed Alfred Wegener and Alexander du Toit to reconnect the continents that originally formed the supercontinent of Gondwana. *Lystrosaurus* left its bones in Antarctica, South Africa and India, along with leaves and fragments of the seed fern *Glossopteris* – highlighting the continuous distribution across lands once connected. Another of Peter's reconstructions was of an ancient diving bird, *Protoplotus beauforti*, whose skeleton had been beautifully preserved in ancient lake

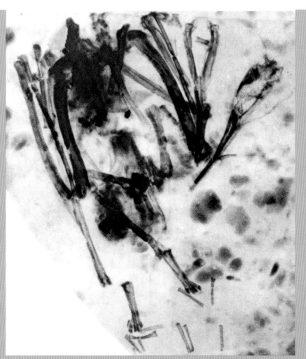

Figure 4.5 An x-ray of *Protoplotus beauforti*, which provided Peter with even more detail of the skeleton of this intermediate bird when he rendered his sketches. The length of the skull is 9.5 cm.

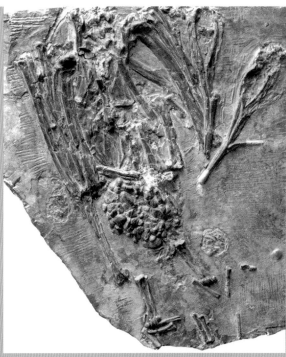

Figure 4.6 The near complete skeleton of *Protoplotus beauforti* held in the collections of the Geological Survey of Indonesia in Bandung. Close up photograph shows the gastroliths, or stomach stones, found concentrated in one part of this skeleton. Such stones likely served both as ballast for diving and grinding stones for processing the food of this active bird (photo by P. Vickers-Rich).

clays of Eocene age (around 50 million years ago) in southern Sumatra.

Protoplotus is one of the missing links, intermediate between the living snake birds (anhingas) and cormorants. I had studied this bird with Rini Marino-Hadiwardoyo of the Geological Survey of Indonesia and CSIRO (Commonwealth Science and Industrial Research Organization) zoologist Gerry van Tets. We had all the inside knowledge at hand from which Peter could work. I had taken detailed stereophotographs and written descriptions and interpretation of this bird during a recent visit to the Geological Survey of Indonesia in Bandung. This specimen was unusual in that it was so nearly complete that Rini and I were able to examine it meticulously, bone by bone. We found *Protoplotus* to have a skeleton with some features very like those found in cormorants, while others were more anhinga-like – it was a real mosaic. The specimen was remarkable in its near completeness, the detail with which we could describe it and the mosaic of characters that it displayed. More than intermediary between darters and cormorants, *Protoplotus* deserved its own family. Even after we had x-rayed the block in which the bones were clearly visible, it remained a bird part way between two quite distinct, yet related living families.

The *Protoplotus* skeleton also told us some other tales. This bird had a gizzard full of well-rounded pebbles, called gastroliths. A wide variety of vertebrates – both herbivores and carnivores – have such gastroliths. Cormorants have gastroliths that probably serve two purposes – ballast for diving and grindstones for processing food. After our detailed comparisons with both cormorants and anhingas, our conclusion was that *Protoplotus* had feeding habits much more like those of cormorants, based on the unspecialised nature of the neck vertebrae and the shape of the skull and beak. It likely fed on small fish, snails and invertebrates, all of which were found as fossils in the lake deposits in which *Protoplotus* was preserved.

And so, although Peter did not actually visit the collections in Sumatra, he had extremely detailed information to work from and our discussions of the evolutionary intermediacy of this form were as researcher to researcher. Peter also had a great asset that I lacked – he was a keen bird observer and had rendered many beautiful and detailed artworks of living birds. He knew well the anhinga and the cormorants, the latter of which were close at hand in the bays in the Melbourne area. It was only years later that I had a chance to get up close and personal with anhingas, when I sailed down the Amazon and met members of this family in a tree! But, at the time we were working on reconstructions for *The Fossil Book*, Peter had the best knowledge of the living offspring of the fossils we were reconstructing. This was a fundamental asset that so many of us paleontologists lack.

Our book used other artwork, some by Frank Knight that had been rendered mostly for another book, *Kadimakara*, as well as a few more works that Peter produced specifically for *The Fossil Book* – sketches of primitive mammals, modified from the reconstructions of others. And our research colleagues from around the world, as well as scientific photographer Frank Coffa at the Museum of Victoria, were generous in providing new illustrations and research data to bring our book up to date. They all shared their research results so that this book could pull the reader into their laboratories, and they expected nothing more than a thanks for this. Their generosity of spirit was most deeply appreciated. Many, like Tom and I, had early memories of the Fentons' book, and I think they looked on taking part in this revision as a true honour. The very special aspect of Peter's art in this book was his attention to detail. By this time, and even from when Tom and I first worked with Peter, his art was not just about us researchers giving Peter instructions and him putting these to paper. Peter was and has always been a collaborator. In the case of *The Fossil Book*, Peter's input was to question and co-develop the concept being created, and he was willing to change it in the future if new finds dictated the need for change. He is also a formidable critic, and unless I was able to document and explain my theories on how an image should be rendered, I had a fight on my hands – friendly, but firm – and after that we usually had a good glass of the best red wine in the house!

Figure 4.7 *'Protoplotus beauforti'* (1985). Stylus pen on paper, 16 x 28 cm, private collection (P. Trusler). *Protoplotus beauforti* from Eocene lake deposits in southern Sumatra represents a mosaic of characteristics between living cormorants and snake-birds, or anhingas. The length of the bird's skull is 9.5 cm.

The Artist: Peter

The provision of a small series of panoramic scenes for the revision of the famous *The Fossil Book* by Fenton and Fenton was my first production in this genre. I say this because I accepted it at face value – I took on the work to correspond to illustrations that had been rendered long ago by Carroll Lane Fenton, and in the same style. The concept of the widely used panoramic format in fossil art has a long history, and I will detail my approach to this style in Chapter 13.

Pat and Tom were working with Mildred Adams Fenton to revise *The Fossil Book* with a wider contemporary coverage. I was simply asked to provide some graphics that were in keeping with the monochrome reconstruction illustrations prepared by Carroll Lane Fenton for the 1958 production. It did not have the lavish, full-colour production of other works that were becoming increasingly common in its day, but it was a scholarly and accessible textbook.

The original illustrations that Carroll Lane Fenton had produced were remarkably consistent in style and level of accuracy for their day, and were amazing for the breadth of the fossil biota that he reconstructed. This spanned more than 3.5 billion years of evolution. The book had withstood the test of time and remained a standard for nearly three decades, in great part due to the large suite of illustrations.

A series of species for each of seven chosen settings was needed for my part in the revision, and I prepared small reconstructions in a simple, animated style. Some of these were more expansive dioramas than the majority of Carroll's work. This was purely for economy. I had chosen a standard stippled technique, in harmony with Carroll's work, even though it was not the same medium that he had employed, in order to maintain some continuity.

The concept of an entire palaeo-community that clearly featured certain key species representative of a particular time, or ones that signified important evolutionary events, was of particular relevance to the new text being written for the revised edition. When Pat took on this job, she had recommended that the new edition should be globalised, and Mildred was happy with this. Therefore, this aspect of significant regional periods in time assumed a greater importance for the small number of new illustrations to be added to the title.

I was to deal with a variety of settings, from the vast freshwater and brackish lakes of the Central Australian Miocene to the rainforests of the Cuban Pleistocene, the Antarctic Triassic as well as the lakes of the Indonesian Eocene and the foothills of the Chinese Palaeocene.

Figure 4.8 'Gondwanan Triassic fauna' (1985). Stylus pen on paper, 17 x 28 cm, private collection (P. Trusler). A Triassic scene from Antarctica or South Africa. A *Thrinaxodon*, a carnivorous mammal-like reptile, guards its recent kill of a lizard-like reptile, *Procolophon*. In the background lurks a crocodile-like thecodont, *Chasmatosaurus*, hoping for some leftovers. On the left are three herbivorous mammal-like reptiles, *Lystrosaurus*. The length of *Lystrosaurus*' skull is 12–15 cm.

Figure 4.9 'Asian Cainozoic fauna' (1985). Stylus pen on paper, 16 x 28 cm, private collection (P. Trusler). A Late Palaeocene landscape (about 60 million years ago) in eastern Asia. In the foreground a small carnivorous *Cimolesta* glances at three pika-like anagalids, while in the background a pair of *Ernanodon* come to drink at the water's edge. *Ernanodon* was once thought to be an edentate like those best known from South America, but it may in fact be more closely related to pangolins. *Ernanodon* was about the size of a sheep.

Figure 4.10 'The Eocene Messel fauna, Germany' (1985). Stylus pen on paper, 16 x 28 cm, private collection (P. Trusler). A Middle Eocene (50 million years ago) forest floor in Germany. A primitive artiodactyl, *Messelobunodon*, forages in the upper left, while others behind it have settled down for a rest. At the right, *Eurotamandua*, an anteater-like animal, is getting ready to attack a termite mound while a closer relative of the living pangolins, *Eomanis*, observes the activity. A primitive hedgehog, *Pholidocercus*, devours a beetle from the decaying log at the lower left. *Eomanis* was about the size of a living pangolin.

The potential diversity of the organisms that could feature in these situations was large. For example, it was ultimately decided to include some of the spectacularly unusual forest animals from the Eocene Messel shales in Germany. This entailed the juxtaposition of primitive artiodactyls with early pangolins, a *Tamandua*-like form and a primitive hedgehog. This was a strikingly unusual combination in modern terms, and this uniqueness factor was to become a common theme in this work.

Despite the simple graphic presentation suited to the revision, one of the main issues for me as the illustrator was to make the characters in each scene appear as a natural part of the play. While the presentation in this genre is understandably artificial, it still needs to convey a natural relationship.

Such naturalism is an artful balance between two elements and the contradictions they contain. Firstly, it must present a visual believability, despite the fact that it is not , strictly speaking, a single, visual image or one photo frame. It is actually a synthesis or composite of many probable images. Secondly, it must present a credible narrative, for this is also a construct of many hypotheses. The events may not necessarily be related or have occurred simultaneously. The art here is one of presenting the entire play in one image, and have it appear sufficiently plausible to allow the audience to examine each act or component independently. The reality, therefore, needs to work at different levels.

With the appropriate setting chosen by Pat, Tom and Mildred, the cast selection for each time frame became a round-table discussion between artist and scientist. And with that ongoing intellectual fencing, the play began!

Figure 4.11 'The Miocene fauna of central Australia' (1985). Stylus pen on paper, 16 x 28 cm, private collection (P. Trusler). Miocene (20 million years ago) forest floor in Central Australia. Three *Ngapakaldia* (about 1 metre long), primitive relatives of *Diprotodon*, have come to one of the permanent ponds for a drink. From the tree above they are observed by the squirrel-like marsupial, *Ektopodon*, while a primitive browsing kangaroo remains quiet in the distant bush. At this time, central Australia was graced by extensive forests, oft-times restricted to river corridors but still providing refuge for arboreal and browsing forms. Only later did aridity take over the landscape, pushing many forms to extinction. The skull of *Ektopodon* reached about 5 cm.

Figure 4.12 'The Early Miocene lake fauna of central Australia' (1985). Stylus pen on paper, 17.5 x 28 cm, private collection (P. Trusler). In the Miocene (some 20 million years ago) permanent lakes existed in central Australia and were inhabited by a wide variety of water birds. Palaelodids (lower right) were similar to those known at the same time from Europe and North America. Several flamingo species made their nests and raised their young in these dependable, somewhat saline water bodies, but this family no longer occurs in Australia. Cormorants, together with a variety of pelicans, rails and ducks, have survived the increasing aridity. Remarkably, freshwater dolphins, perhaps most closely related to the platanistid river dolphins of the Ganges River today, also once occurred in some of the more southern lakes near today's Lake Frome in South Australia. The largest flamingoes reached the size of the living Greater Flamingo.

Figure 4.13 '*Ornimegalonyx*' (1985). Stylus pen on paper, 14.5 x 28 cm, private collection (P. Trusler). *Ornimegalonyx* was one of the largest owls and is known from the Pleistocene (about 1 million years ago) of Cuba. Its wings were much reduced, which indicated that it did not rely on strong, powered flight. It is attacking a nearly 60 cm long insectivore, a *Solenodon*.

5
Dinosaurs from China

Figure 5.1 Poster announcing the opening of the Dinosaurs from China Exhibition in Melbourne, 1982.

The Scientist: Tom

During 1980, I attended a small reception for a few visiting museum scientists from China. An entomologist from the delegation, on realising that I was a vertebrate palaeontologist, immediately dug out of his pocket some photographs of specimens housed at the Tianjin Museum, where he worked. The first was of a mastodon. I responded to it with a polite nod. The second was of a skeleton of the Late Jurassic sauropod *Mamenchisaurus hochuanensis*. Again, I initially responded with a polite nod. But then the entomologist said in his best English, '… and this is going to Japan'. At that remark, my interest shot through the roof. 'If the specimen could leave China and travel to Japan,' I thought, 'it has to cross water. And if it can cross water, surely it can go just a bit further, like to Australia?'

I hurried to introduce the entomologist to Barry Wilson, my director at the National Museum of Victoria. The entomologist repeated that, yes, this giant dinosaur specimen was going to Japan, as he politely showed Barry the same photograph.

Could Barry and I persuade the director of the Tianjin Museum to allow us to borrow the *Mamenchisaurus* skeleton, subsequent to the exhibition in Japan? As he and I thought about writing the official letter direct to the Tianjin Museum, an even better idea occurred to me.

It just so happened that Pat was about to make her second trip to China, specifically to the Institute of Vertebrate Paleontology and Paleoanthropology (IVPP) in Beijing. She had become a research colleague and friend of several staff

Figure 5.2 Skeleton of *Mamenchisaurus* in Tianjin Museum, China. The skeleton is more than 20 metres long.

members there on her first visit in 1979, and was to return in 1981 to continue her projects. She was working on the compilation of a Chinese–English/English–Chinese dictionary of palaeontological terminology with Prof. Minchen Zhou (or Chow) and his contacts throughout the People's Republic of China. Perhaps, if she carried the letter intended for the director of the Tianjin Museum with her to China, it would give us a better chance of reaching him – more so, if it was then conveyed by one of his own colleagues, Minchen himself. So, Pat personally handed the letter to Minchen. His reply was quick and most encouraging: 'Actually that specimen belongs to IVPP, and I shall help you get it.' We knew then that the mooted exhibition had a good chance of becoming a reality, the first major dinosaur exhibition to ever come to Australia.

So, the deal was done, but before the fossils could leave for Melbourne a formal contract between IVPP and the National Museum of Victoria had to be agreed on and signed. Prof. Jim Warren of Monash University, a senior member of the Board of the National Museum of Victoria, flew to Beijing to negotiate that agreement on the

Figure 5.3 Pat with staff of the Institute of Vertebrate Paleontology and Paleoanthropology in Beijing, 1979. In front are Minchen Zhou, Pat, Sun Ai Lin and Wang Banye. Banye also worked on the dictionary project with Pat, and along with Zhang Yu-ping travelled around China with Pat on her first trip to the People's Republic of China (photo by Zhang Yu-ping).

Museum's behalf. Just prior to his departure, I casually raised another matter with Barry Wilson and Jim: 'As the Chinese seem ready to loan the specimens to us for display, why don't we ask them if they would also agree to our moulding the two largest dinosaurs, on the understanding, of course, that they and we shall each get a cast and they shall retain the moulds in the end?' When Jim returned a few weeks later, he had quite successfully negotiated an agreement to borrow *and* mould the two largest dinosaurs, the sauropod *Mamenchisaurus hochuanensis* and a hadrosaur, or duckbilled dinosaur, *Tsintaosaurus spinorhinus*, and make two casts of each.

Within a remarkably short time, the skeletons of these large Chinese dinosaurs became the centrepieces of the *Dinosaurs from China* exhibition at the National Museum of Victoria. These were the first complete, real, and very large dinosaur skeletons to be displayed anywhere in Australia. As a consequence, the attendance broke all records at the Museum.

Technicians and scientists from IVPP accompanied the collection to Melbourne. They helped set up the skeletons of *Mamenchisaurus* and *Tsintaosaurus* for our display, as well as mould and cast them along with our own technicians, Mike Traynor, Peter Swinkles, Jim Couzens, Dean Smith and others. Moulding and casting began about six months before the display opened in Melbourne and continued while the exhibition was underway, both in Melbourne and subsequently at the Australian Museum in Sydney. Because this ambitious project involved not only technicians from the National Museum of Victoria, but also a number of visiting scientists and technicians from China, it provided a unique opportunity for interaction between people that had been separated from each other for decades – East met West, or should I say North met South!

Peter was commissioned to produce illustrations of the skeletons of *Mamenchisaurus* and *Tsintaosaurus*. He sketched individual bones that lay scattered about on the museum floor as they awaited moulding. He then integrated these sketches in a life-like pose, both as stippled diagrams and watercolour monochrome paintings. The watercolour painting was subsequently used both for posters announcing the exhibition and in the exhibition catalogue that Pat, Erich McClellan and I wrote – our first.

In the years that followed, the casts of *Mamenchisaurus* and *Tsintaosaurus* were mounted and remounted a number of times, first moved about within the McCoy Gallery of the old National Museum of Victoria on Russell Street, then twice within the newly built Melbourne Museum at Exhibition Gardens, where they now reside. (This museum now operates as Museum Victoria.)

Wondering if it was plausible for *Mamenchisaurus* to stand on its hind legs with its neck upstretched, as a few artists had previously rendered restorations of sauropods, I examined its pelvis. What particularly struck me was that the part of the articular surface where the femur would contact it when the animal stood upright was thicker than the surface of contact when the animal stood on its four feet. This made me think

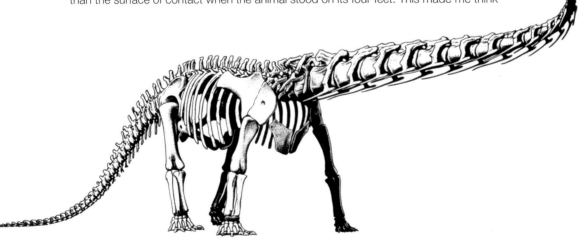

Figure 5.4 *'Mamenchisaurus* – skeletal reconstruction' (1981). Watercolour on paper, 28 x 38 cm, IVPP collection (P. Trusler). Skeleton of *Mamenchisaurus hochuanensis*, more than 20 metres in total length.

that the pelvis was capable of withstanding a greater stress when the animal was upright, and thus it was indeed plausible for *Mamenchisaurus* to stand that way. So, in 1989 it was remounted in the McCoy gallery in that pose – a stance which has since been used for other sauropod skeletons in museums around the globe.

When the cast of *Mamenchisaurus* was originally put together, its skull was yet unknown. So, we used the skull of another sauropod as a substitute. In 2002 the skull of a second species of *Mamenchisaurus* was discovered and described. Working from the illustrations in a research paper, Tom Davies, one of the technicians at Museum Victoria, then produced a life-sized model of the skull, which replaced the substitute skull on our cast of *Mamenchisaurus* when it was remounted in Museum Victoria.

The *Dinosaurs from China* exhibition was a milestone personally and professionally. It initiated long-term research with our Chinese colleagues. It also set in motion cooperative links between our institutions and several others in the People's Republic of China. Pat continued her research there in the 1980s and 1990s, and we still work with researchers from IVPP on projects related to Cretaceous mammals and the palaeontological dictionary. Finally, as we write this in 2010, the remaining vocabulary collected by Pat since the early 1980s is about to be published in the second of two dictionaries of palaeontological terms. Both will now be put online for future use.

Students and researchers from our respective institutions have moved back and forth between these two nations over the intervening years, beginning in the 1980s: Patricia Komarower completed a dissertation on the history of IVPP; Zhang Yu-ping from IVPP worked on the dictionary project in both Beijing and Melbourne; Fan Jun Hang from IVPP came to Monash University to pursue a PhD; Zhou Siqing from Beijing was an artist in residence at the National Museum in 1982 – just a few of the outcomes of this cooperative interaction. Many further ventures developed from this first contact in the late 1970s.

The *Dinosaurs from China* exhibition was also the beginning for Pat as a builder and director of major international travelling exhibitions – ventures that saw research scientists presenting their results to the public, bringing visitors to these exhibitions into the scientists' world. And, as an aside to this, these same researchers used such public exhibitions as funding generators for underwriting research, supporting students and linking up international cooperation with other research groups, in a time of diminishing resources. Importantly, the generation of the resources and the funding has been a central component in facilitating Peter's work. He has continued to freelance, and no salaried position for him as an illustrator has ever been created.

Peter's original art pieces and the moulds produced from the real bones also returned to IVPP, as agreed on from the beginning. More casts were poured from those moulds, and Peter's reconstructions of both *Mamenchisaurus* and *Tsintaosaurus* have been widely used around the world in many books, exhibition catalogues, posters and a myriad of other forms, to this very day. Both images from this first dinosaur exhibition to ever visit Australia have become global icons.

As the title of a little booklet published by Australia Post much later (in 2008) about Australian megafauna so aptly noted, *From Small Things Big Things Grow*.

The Artist: Peter

During my time illustrating the *Diprotodon* specimens from Bacchus Marsh, I was only vaguely aware of the other events unfolding, in the form of occasional letters and phone calls which passed Tom's desk. I was not cognisant of the international diplomacy that had been transpiring, for although I was aware of the technical dictionary interests that had taken Pat and her colleagues to China, I was not initially a confidant of their grand exhibition plans. But, they clearly had me in their sights for this too, and it was soon to impact on me.

A vast array of specimens and ancillary material was to arrive for the show. The exhibition included phenomenal specimens, the likes of which I had not seen before. I had never chanced on the names of most of the 'stars' in the show, so, in this sense, all was new to me. My experience with the vast collections of Mesozoic faunas from the northern hemisphere was essentially confined to books. So, when I first laid eyes on the huge crates of petrified bones being unpacked and dispersed across gallery flooring by a small army of storemen and packers, scientists and preparators, I was speechless. The size and weight of the real thing has to be touched and tested to be believed. Even so, the remains of such creatures are barely comprehensible to me to this day.

It was certainly a privilege to be involved, although the small illustration task that I was asked to undertake was essentially to provide some promotional graphics for the exhibition. Of itself, this was no big deal, despite a certain pressure resulting from the significance and timing of the event. The truth of my circumstances was one of acute self-consciousness. I was dumb-founded by the enormity of the primary specimen material and aware of my lack of knowledge about dinosaurs in general. I had been interested in the revolutions of understanding that had been taking place in dinosaur science throughout the 1960s and 1970s, and now something that was, in a sense, academic curiosity was directly confronting me. I could have approached this purely from a graphic design point of view and simply transcribed or amalgamated imagery from the suite of photographs that supported the Chinese collection. There was a complex of reasons that dissuaded me from this easier option. Sure, there were time constraints and access issues because of the moulding and casting of the fossil elements that were going on at the same time. There were ongoing scientific and technical discussions about the details of the mounting of the material, and the finished art was required in advance of the final assembly of the specimens. There was a cultural dimension transpiring as well, and this interested me, because it was manifested via two sources.

The first of these cultural interactions was in the person of a young artist like myself, Zhou Siqing, from Beijing, whom I had got to know at this time. I witnessed the production of his work while he was an artist in residence at the National Museum of Victoria in Melbourne. We occasionally discussed his responses to and comprehension of things Australian. These influenced him personally and were expressed in the progressive changes to his art practice and the subjects capturing his imagination. The second dimension of culture was the variety of visual material that came from China to support the show. Both revealed conflicting issues of influence and assimilation – sometimes producing astounding outcomes and in other instances poor ones. Siqing himself was straddling ancient traditions and recent cultural changes occurring in Maoist China, just as much as he was confronting the modern Western culture characterised by Australia. His acute inquisitiveness about the rapidly changing dialogue here with respect to Aboriginal culture highlighted the most fundamental nature of all differences, and the adjustments that were being made in my own culture. It was interesting to see the cultural strengths from both countries mix in this way with each of us already, and simultaneously, experiencing pivotal internal changes – all the more so because on this occasion it was occurring in the realm of a scientific pursuit rather than a purely artistic one. There were cultural differences to be seen in the science too, and I think none of us foresaw the global magnitude of the exchanges this event was to herald.

Figure 5.5 Block print by Zhou Siqing of Australian fauna. Siqing spent one year at the National Museum of Victoria as an art scholar.

It seemed, therefore, that I could, and should, contribute much more than a simple sketch. But first I had some homework to do! I needed to tease out the current scientific consensus and controversies which might apply to the two dinosaur species that were to feature in the exhibition. I needed, of course, to understand the characteristics of the two disparate groups of dinosaurs they represented, one being a sauropod and the other an ornithopod. Furthermore, there were specific features for these Chinese forms that were new to science. This was not entirely the quiet task of thumbing through the pages of weighty references, nor was it a case of trying to recall lectures on Mesozoic evolution from a few years previously. There was considerable, and ongoing, debate on all of these matters, constantly filtering from conversations across the gallery as I sketched the bones at my feet. With the aid of the field photos from the respective excavations of these articulated skeletons, and taking note of the way these two skeletons had been mounted in Beijing, I formulated a basic posture for each. All the while, Tom cast an occasional glance over my shoulder. I amassed sketches for the individual bones and partly articulated components from perspectives that applied to their place in the proposed postures. This would allow me to compile a more detailed rendering of the articulated skeletons than any of the previous photos. It facilitated my familiarity with the physicality of the fossils – that is, I actually looked at the detail of the bones and articulation between them – which would ultimately allow us

Figure 5.6 Skeleton of *Tsintaosaurus* at the National Museum of Victoria, Russell Street, just prior to the launch of the *Dinosaurs from China* exhibition. With Brent Hall, preparator, for scale (photo by F. Coffa).

Figure 5.7 '*Tsintaosaurus* – skeletal reconstruction' (1981). Watercolour on paper, 28 x 38 cm, IVPP collection (P. Trusler). *Tsintaosaurus* was the second large dinosaur skeleton in the *Dinosaurs from China* exhibition. Peter's reconstruction drawings of this dinosaur and *Mamenchisaurus* prepared for this exhibition have had widespread use in many other exhibitions and projects. After the travelling exhibition returned to China, the moulds from which the casts of these two dinosaurs were pulled to grace the exhibition hall at Museum Victoria were returned to the Institute of Vertebrate Paleontology and Paleoanthropology (IVPP) in Beijing, along with Peter's original illustrations. Figure 5.6 shows the skeleton and gives an indication of scale.

to fine-tune any matter of postural and lighting change. The use of the small illustrations that I was to prepare was going to include large scale reproductions for banners and so on – this art needed to cater for extreme enlargement, and considerable reductions for merchandising purposes as well. To be on the safe side, I produced two stylistically different versions.

With the science, the exhibition, and the meeting of cultures as the backdrop, all seemed dynamic, and so I decided to exercise a little licence for the reconstructions. For the sauropod *Mamenchisaurus*, its gargantuan disproportions were always going to present illustration headaches. It had an absurdly long and well-buttressed neck that really had the scientists scratching their heads over its postural implications and its flexibility of articulation. (Interpretations of these were to change again following the wide exposure that this exhibition was to facilitate.) Ungainliness was, in some respects, *Mamenchisaurus's* most intriguing asset! This generated great difficulties in making a static skeleton lively as well as interesting to a wide audience.

Everything about *Tsintaosaurus* was much more amenable to creating an illusion of life in the presentation of a skeleton. While the degrees of flexure of the tail and neck were debatable – certainly by today's understanding – the low angle of view here provided sufficient perspective illusion to enhance its animation without overly distorting the skeletal proportions. A certain elegance of mobility could be achieved and a sense of consciousness could be inferred by presenting a posture as if the dinosaur had just momentarily been distracted and raised its forequarters in order to obtain a better view.

As a graphic, the two piles of bones that lay before me on the Museum floor needed to assume a personality that worked instantly, hopefully without me having to resort to either kitsch or cliché. To me, these skeletons had to have a sense of scholarship befitting the science of the time and, importantly, lend a cultural flavour to the entire presentation suggestive of 'East meets West' as well as a sense that 'old' transpires to 'new'.

These were unusual considerations to be applied to extinct organisms from some 150 million years ago. Whatever anthropomorphic improprieties I may have exercised in this task, it did serve me well in one important respect. I became aware of just how subtly I could potentially infuse quite simple illustration with all manner of unrelated emotional baggage. There were all manner of human considerations that could be directly applied to, or indeed, unconsciously transferred to, illustration at any level. These could actually be used to ensure the purpose of the work was achieved by making it look comfortable, appropriate or correct. This invariably carries a very subjective component and an intuitive knowledge of normality, as opposed to actuality. And when it comes to illustrating for the advancement of science, or culture for that matter, it is commonly independent of the facts and counter-productive. This is not to say that it cannot be done with good purpose. But without acute awareness, usually it is detrimental, stifles advance and most often affects the production and appreciation of work at a totally unrecognised level. Furthermore, it can be the stuff of whimsy and mischief.

From an artistic standpoint this emotional element represents power and flexing this power can be highly creative. From an illustrative position I can potentially make it positive or negative with respect to the purpose of the artwork. This assumes that I have a great deal of cognitive control, which I do not have, and it places a great demand on both myself and the trust that my colleagues place in me. Indeed, have I the moral conviction to exercise an honest adherence to facts for the purposes of scientific illustration? It would presume that when I invariably do employ creativity in the scientific process, it is done with respect to advancing understanding, and that must be honoured above my own personal aspirations or those of my colleagues. These are all value judgements, and there seemed no easy way through.

In relation to dinosaur skeletons and with my own dilemmas in mind, one simple example will hopefully illustrate these powerful forces at work in a seemingly benign and factual world.

As the scientific community started to revolutionise understanding of the metabolic and functional capabilities of these Mesozoic giants, dinosaurs suddenly got active, very active. There was a remarkable awakening in the 1970s and 1980s, and artists everywhere looked for ways to incorporate this new knowledge of anatomy and biomechanics in presenting the research – in bringing these new ideas into a wider appreciation of the dynamic history of our planet. As I looked for ways to enliven information presentation with this new understanding, everyone else was doing much the same. The best of these practitioners have taken pains to minimise the infusion of cultural mythologies into their scientific art at a professional and at popular level. But for the moment, one small thing has become stuck, seemingly formalised as an unconscious protocol for all dinosaur skeletons, bipedal or quadrupedal.

Just as one can make an outline drawing of a machine appear to move at high speed by forward sloping its vertical lines, this psychological animation trick has essentially been applied to skeletons. It shifts the apparent centre of gravity and infers that the static object is likely to move in that direction – a combination of the experience with unstable objects and our own proprioceptive body workings. The physical laws apply to animal locomotion too, and the style of movement and body form are highly variable in its instantaneous expression. The body conformation at any one 'frame' of motion depends greatly on the skeletal proportions of the body and limbs, the number of limbs being employed, and the type of gait of the animal, as well as the speed and the phase of the stride being depicted. These matters are well known to biologists and sports physicians alike, but the current scenario to which I refer is a little like the pre-photographic depiction of horse locomotion as opposed to the reality that photography later provided.

Palaeo-science has had a revelation, but illustrators have only provided part of its expression. Once the idea occurred of thrusting the dinosaur's centre of gravity forwards by placing the pectoral and/or pelvic limb pairs with one leg fully retracted and the other fully extended posteriorly, the beast was perceived to be able to run.

All issues of appropriateness, hyper-extension, biological capacity, gait variation, and so on seem to have been abandoned for the illusion of speed. So universal has this convention become in palaeontology that throughout scientific literature in recent decades, and despite the depth of study being presented, dinosaurs no longer stand; they are almost without exception depicted as running at full speed! Irrespective of the purpose of the illustration or diagram, every one appears like a human sprint athlete's skeleton, which has just left the starting blocks on the tips of its toes. It is very anthropomorphic, looks highly convincing, and is probably wrong in most cases. Such formalisation hopefully serves to highlight the need for artists and scientists alike to be considerate of both their subjects and conventions, because the staggering diversity deserves individual attention.

Formalisation is a significant part of both recognition and acceptance. It plays a part in reconstruction art and is really the antithesis of inquiry, because it employs cultural and human behavioural cues. How much might a simple psychological cue reinforce the status quo? What is the context for the production of the artwork, and does it matter? This small project, above all, taught me about something I jokingly call 'visual literacy'. In order to express and communicate understanding of the past through visual means – especially a past without human presence – do I need to make it more human? To me, the answer is a resounding no, and yet alternative means are not obvious. 'Literacy' in this sense is a capacity to view imagery produced via analytical thought and to be able to perceive in it, or read from it, those components of humanity that are confounding the message from a greater understanding. Artists actually need this literacy to become good illustrators.

Perhaps a simple analogy might make the point clearer. In motion picture animations of large predatory dinosaurs, have you noticed that the vast majority of them roar and growl like lions or bears? There is not one shred of evidence to indicate that they made such noises. If their nearest reptile or avian relatives are anything to go by, they hissed, squawked, cried, twittered or warbled – anything but deep roars and howls, which are the aggressive or territorial vocalisations of large mammalian predators. The roaring concept has been unconsciously or deliberately invoked to fit a human psychology, because in our modern world these are the natural (animal and non-human) sounds that most easily initiate a primal fear. Few of us have ever been in a context to genuinely have the psychology confirmed, but it is instant and innate. If the ancient beast is to look fearful – and they do to our eyes – then surely their vocalisations must be fearful too? You can see the weak logic here, and its ramifications, and yet it is a cultural and scientific fallacy that has persisted almost since the dawn of the motion picture industry. Only recently have there been some attempts to communicate otherwise, or desist from the obvious misconception.

The same scenario is equally profound with imagery, and the transference of human anatomical features and even human sexual and social psychology abundantly appears in 'palaeontological art'. What is needed is a more sensitive use of this tool by us artists and a correspondingly greater critical awareness or discrimination in the scientific and viewing public. This is something I have striven to develop in my own work, and Tom, Pat and I have correspondingly sought to encourage critical perception in our audience, especially students at any level.

Figure 5.8 In 1982 the first travelling dinosaur exhibition opened at the National Museum of Victoria (now Museum Victoria) on Russell Street in Melbourne. Peter's work graced the banner welcoming visitors into the Aussie first.

6
Wildlife of Gondwana

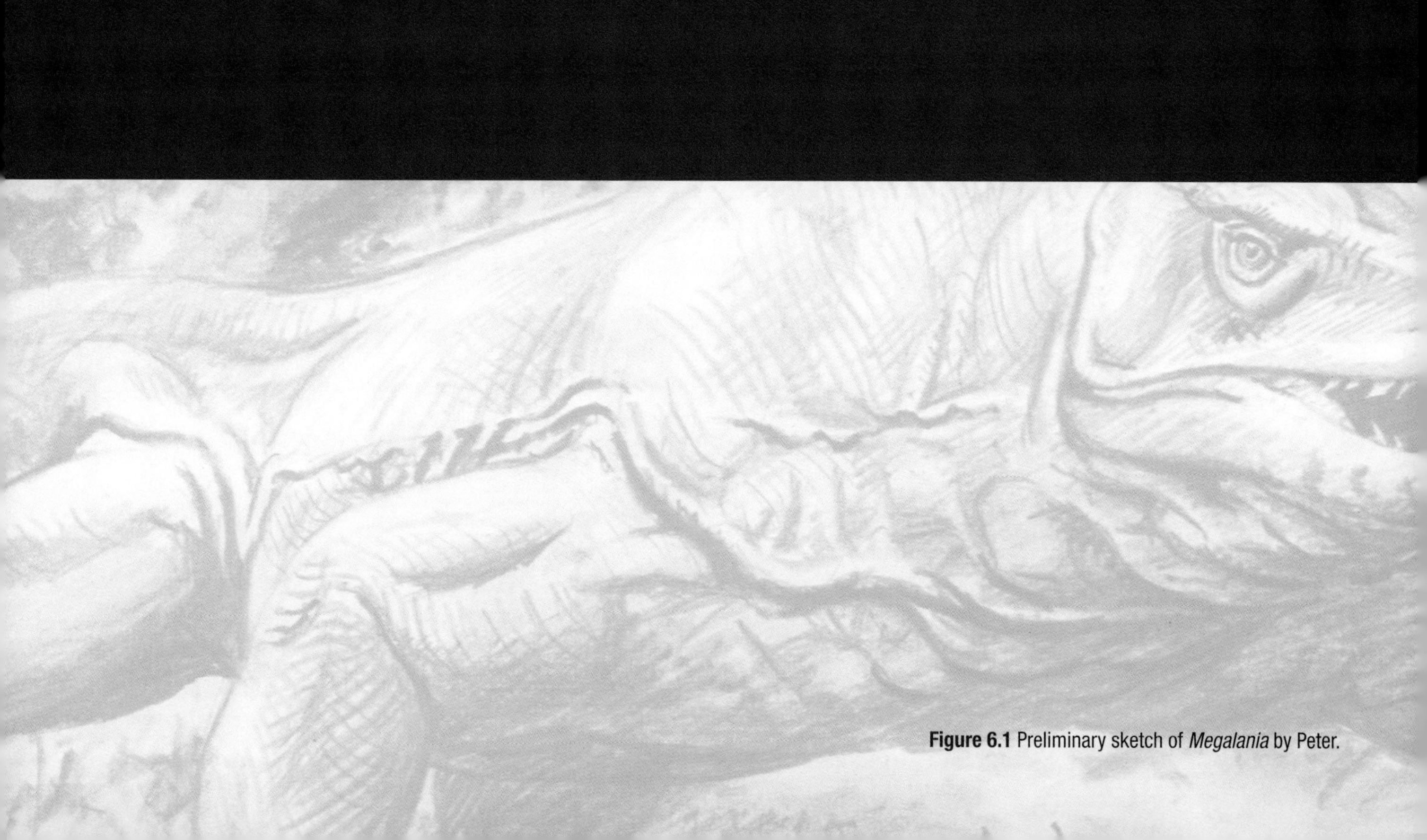

Figure 6.1 Preliminary sketch of *Megalania* by Peter.

The Scientist: Pat

*'Time present and time past
are both present in time future'*

T. S. Eliot

Friday, 30 August 1985. Looking out at the audience gave me pause to think. This was Dallas Brooks Hall in Melbourne (Australia) – a presentation organised for students in conjunction with the international meetings of ANZAAS (Australia New Zealand Association for the Advancement of Science). The organisers thought that in parallel with this professional conference it would be appropriate to run a youth conference – and this conference was significant enough that Sir Edmund Hillary, the first Westerner to top Mount Everest, had been selected as the patron.

The kids were restless – more than 1500 of them on school break and massed here to listen to scientists tell them about what they did with their lives. I wondered how our presentation would go over. The colourful sea of chattering students, some in school uniform, were definitely at an age of challenging authority – would they listen? Would this mob learn something from what we had to say?

One of my kids was on school break, too, and in late June of this year I had given birth to a son, so there was also a mother's time management issue that I had to have under firm control. My 8-year-old daughter and her tiny brother Tim had come with me to this conference, and were sitting in the parking lot in our aged Land Rover with the babysitter, Faye, quietly knitting, fully in charge of the kids.

So, there I was, poised on the stage of the large concert hall with a variety of management issues and some very unusual companions. I have always enjoyed having no rehearsal for public presentations, and this was no exception. One of the facilitators of this morning's event, David Smith, had dressed himself in a rabbit suit and was cruising around the stage and the hall preparing for the event to get moving. I was never quite sure how he would introduce us, so, like it or not, this action had to be *ad hoc*, and good, if we were not to lose our audience. Our presentation was to be about time – a long time spanning back billions of years – and in the spirit of T. S. Eliot, we were trying to make the point that there were lessons to be learned from the past that informed how to proceed in the future.

Figure 6.2 Footprint of a dromornithid related to *Genyornis*, found in Cenozoic sediments near South Mount Cameron in northeast Tasmania. Many footprints of these ancient birds were preserved in this region, giving some idea about how they moved (photo courtesy of Robert Green). The footprint's length is about 20 cm.

There were three of us on stage – well, actually four, or really five when I think about it. The most impressive was the skeleton of a big lizard, about 5.5 metres (18 feet) long – *Megalania prisca,* reconstructed as moving towards a somewhat smaller skeleton of a big bird, *Genyornis newtoni*. Sitting quietly beside the predator and its prey with me were two human companions – and the 'rabbit' was darting in and out, around this mosaic of humans and beasts.

One of the humans was technician Craig Cleeland, an undergraduate at Monash University, in the then Department of Earth Sciences. His undergraduate project had been to mould the bones of the extinct, giant ground bird *Genyornis* and produce a cast for this project, with the cast to be later used at Monash for teaching. (Now in 2010, that cast lies in the National Museum collections in Timor-Leste, where it will one day form part of their permanent displays on the history of Timor and Australasia – but that is a story for another time.) Craig had also been in charge of assembling the cast of a companion for *Genyornis* – a giant goanna – and had done this in his father's driveway. This had, understandably, attracted the attention of the local newspapers.

Also sitting on stage beside me was Peter Trusler. Peter had been engaged in this project to produce a series of art pieces which clearly told the story of how a palaeontologist works from the bones to provide reconstruction of the complete animal, in as careful and detailed a fashion as humanly possible. The plan for this presentation was for me to talk first about how the bones of *Genyornis* had been discovered, dug up, and ended up in the South Australian Museum in the late 1800s, and then how they were prepared and studied. Craig was then to take over and discuss how he had produced the cast of *Genyornis* from the real bones, how he had estimated the missing elements and how he assembled the skeleton. Peter was to pursue the path he had gone down to reassemble the bones in a dynamic pose – not just standing there, eyes forward, completely lifeless and boring. Peter had decided to strike the pose of a bird being startled off a nest of eggs by a very large, very predatory lizard. *Megalania* was one of the main carnivores in the Australian Pleistocene, along with truly terrestrial crocodiles, and a few smaller marsupials with a taste for meat rather than salad!

Peter was to explain how he used his observations of the modern world to breathe life into these long-dead bones. This meant initially adding muscles as indicated by the bumps and ridges on those bones, which he could understand on the basis of his observations and dissections of living lizards. Then he had to cover all that with skin and come up with a good reason for the colour patterns he placed on his final image. This certainly meant examining illustrations of living relatives of *Megalania*, in particular the Komodo Dragon of Indonesia. I remember watching parts of a TV documentary again and again with Peter, who spent a good deal of time following these giant goannas about their daily lives, checking out muscle actions, behaviour patterns – everything one would like to know about these impressive reptiles, and probably more!

The audience was attentive, the talking and rustling stopped and the rabbit hid behind the stage curtains.

That was the beginning of the art that formed the basis of our joint book *Wildlife of Gondwana*. It won the 1993 Eureka Award for the best book on Australian science. It was also the beginning of my involvement with directly communicating science to a general audience. I was determined that this sort of communication was not to be a dilution or 'dumbing down' of research results, but an engaging presentation of scientific research, albeit with a minimum of technical terms, but keeping the marvellous detail. I continue to be convinced that children, and the public in general, are quite capable of understanding complex science as long as the terminology of explanation is not so technical that it is far outside their everyday vocabulary. They are quite capable of perceiving the conclusions for themselves once given a roadmap of how to get to those end results, those scientific hypotheses. It is, however, more difficult to convince that same public that ideas can change with the gathering of more data, and a robust theory must be constantly tested. Then, if conflicting data come to the fore, that original idea needs to be changed. Science is a constantly changing panorama. It is not a belief system set in stone.

Wildlife of Gondwana was first published in 1993 by Reed Books in Australia, and full credit should be given to Mary White and Bill Templeman for the form in which it appeared. Mary White, a brilliant palaeobotanist, had previously published the beautiful *Greening of Gondwana* with Reed Books, and Bill had managed its production. I liked this book so much that I contacted Mary to ask her advice on producing a companion volume on the animals that had lived with 'her' flora through time. She was delighted and encouraged Reed to pursue this concept.

Reed was absolutely supportive of Peter being the reconstruction artist for this project and provided the additional funding for his work – in advance. This helped not only with production of the book, but also with the preliminary sketches and the research that went into the rendering of two more detailed works (the Devonian Gogo reef in Australia and the corpse of the dinosaur *Leaellynasaura* from the Cretaceous) that together formed the centrepieces of *Wildlife of Gondwana*. Added to the *Megalania* and *Genyornis* art, these formed brief snapshots of Australia over time, ranging from 350 million years ago to around 50,000 years ago – one vision from each era.

Peter continued to liaise and work directly with a series of researchers in rendering these reconstructions, making sure that every aspect of the art was as true to science as possible. In the *Leaellynasaura* painting, for example, such detail as the angle of light at a particular latitude on Earth at a particular time of day was taken into account. Australia at around 106 million years ago lay much further south, at upwards of 70–80 degrees south latitude. At that latitude, day length could be determined and sun angle at a particular time of the day and at a particular time of the year estimated. That is clearly illustrated in the reconstruction of *Leaellynasaura.* There is evidence of ice being present at this time, and so Peter placed this along the side of the oxbow lake near where *Leaellynasaura* died. Why an oxbow? Tom had mapped the site where the skeleton of this dinosaur was recovered, and his interpretation was of such a formation. Peter even made sure that the grain size of the sediments on which he laid the leaellynasaur was the same as that in the rocks entombing the fossil.

Peter chose to paint the leaellynasaur in a position as close to how this partial skeleton was found, rather than reconstructing this saurian as a living reptile. The detail of this corpse was the best interpretation of the available evidence. Further preparation in future might require some alterations to the painting, since science does march on with each new discovery. In fact, the *Megalania* in Peter's original painting had to be modified during production, when more bones of this rare vertebrate were found. That alteration required putting in place a small crest along the midline of the top of the skull, and such modification will likely continue into the future as more and more fossil material comes to light.

Wildlife of Gondwana was a project that involved not only Peter, Tom and me, but a number of our colleagues as well. Many generously opened their collections to allow Frank Coffa of Museum Victoria and Steve Morton of Monash University to photograph their research material in exquisite detail, despite some of this material being as yet unpublished. I was able to raise the funding to send these skilled and imaginative photographers to collections around the country – the Australian Museum and the University of New South Wales, the Queensland Museum, the Northern Territory Museum, the South Australian Museum, the Tasmanian Museum and the Queen Victoria Museum and Art Gallery. Similarly, when the second edition of *Wildlife* was prepared, Frank Coffa travelled to Argentina to photograph a wide selection of fossils from research collections there. Again, many of these specimens had never been described. The first and the second editions of *Wildlife*, by virtue of this attention to detail, became not only books for the general public and fossil enthusiasts, and a classroom text in some instances, but also a valuable source for other researchers.

The growing collection of both the Trusler art and the high quality photographs of Coffa and Morton was becoming an important source of images for travelling exhibitions, which continue to tour globally to this day. The first of these was the *Great Russian Dinosaurs* exhibition, which made its debut in Melbourne in 1993, in the same year as *Wildlife of Gondwana* was published. Had it not been for the art that Peter already had in hand for *Wildlife*, as well as Frank's photographs of the Russian fossils that featured in the exhibition, it would have been impossible to produce the catalogue

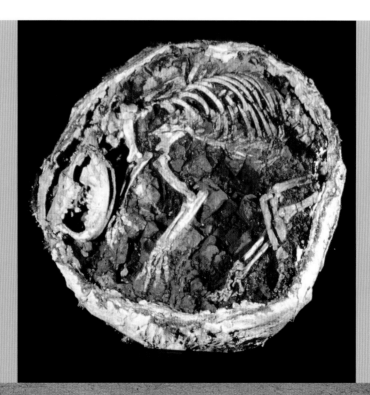

Figure 6.3 *Arminiheringia*, a borhyaenid marsupial carnivore known from Early Eocene rocks in the Salta Province of Argentina. These animals took the role of the Australian marsupial Tasmanian Tiger, and were convergent upon placental carnivores like wolves, and about the same size (photo by Frank Coffa, courtesy of Heime Powell, University of Tucuman).

for the *Great Russian Dinosaurs*. That exhibition opened in August, only three months after the contract was signed between the Paleontological Institute in Moscow and the National Museum of Victoria (now Museum Victoria). There would have been insufficient time to have commissioned such involved imagery to feature on the exhibition panels. Fortunate, too, that Australia Post had commissioned Peter to render the images used for an Australian dinosaur stamp issue that they launched almost simultaneously with the *Great Russian Dinosaurs* exhibition. The gracious involvement of that organisation allowed these images to be used in the exhibition graphics and catalogue. One of these images, of *Leaellynasaura* hatching from its egg, made the cover of the southwestern Pacific edition of *Time* magazine on 9 August that year. At the same time, Steven Spielberg launched his *Jurassic Park* movie – dinosaurs were white-hot, and unintended cross-promotion brought huge crowds to each of these events – movie, stamp launch, exhibition opening. Even the Australian government participated, with the then Minister for Foreign Affairs, Gareth Evans, personally launching the *Great Russian Dinosaurs* exhibition.

Peter's images rendered for these many projects have long educated and fascinated; but, in addition to that, because they are cared for by the research scientists themselves, they continue to generate research funding and funding for further reconstruction art. Most importantly, the reconstruction images produced by Peter have set a high standard for palaeo-art, making clear that detailed art can not only engage the non-researcher, but can also serve as focus for further scientific debate and discovery. *Wildlife of Gondwana* has certainly been such a focal point, and has nurtured many projects not even imagined when *Wildlife* was conceived.

Figure 6.4 Peter's art graced the cover of the Australasian editions of *Time* magazine in August 1993.

The Artist: Peter

This commission crept up on me stealthily. 'Could you reconstruct a Central Australian Pleistocene scene for me to show how a palaeo-artist might progress through the various stages?' This was the gist of a request from Pat. It was nearing the end of 1984. She said she had two animals in mind, and that the scene needed to be suitable for a video presentation to a secondary school audience at a national science conference, ANZAAS, the following year. Animation was not my interest, nor did we have the funds, equipment or expertise to produce it. I think she just threw the idea open to see what I might provide. Science education at all levels is her lasting passion, and I was to become accustomed to her presenting me with an alarming diversity of projects.

This was going to be interesting, although I could not quite come to terms with having to present it in person when she later explained this aspect. I was far more content on the other side of the drawing board, but the prospect of having to formalise the process had its appeal. I greatly enjoyed the working stages of the reconstructions that I had done to date, and so few people ever saw or understood the drawings made en route to the final image. One measure of the success of a painting is that it should appear effortless – full of energy, but not toil! The problem with this project was that it was to be full of complex effort!

Would kids be interested? I was unsure, but Pat quietly insisted that it was to be 'first rate', as she started feeding me with all manner of technical references and a wide spectrum of scientific papers. These hints were not subliminal! The science needed to be well understood, for I was to play an active role in the reconstruction of a skeletal model and cast. These were awaiting articulation in order to conform to the posture and behavioural narrative of the palaeo-scene that I was to create, rather than *vice versa*, as was usually the case for me. The scene was to feature a long-known Pleistocene bird, *Genyornis newtoni*. This was a flightless species, larger than the modern emu, perhaps 30 percent taller, depending on posture, but more robust at an estimated three to five times the weight. *Genyornis* was the last of a long evolutionary line of birds colloquially known as 'mihirungs', or scientifically as the family Dromornithidae. They had been endemic to Australia and subjects of Pat's own PhD research.

The other character of the pair to be reconstructed was *Megalania*, or *Varanus priscus* as it is now known. *Megalania* was a large monitor lizard and probably the largest 'goanna' to have lived on the Australian continent. The monitor group, or family Varanidae, is widespread throughout Asia and Africa and appears to have an equally widespread and long fossil record that includes many large species. *Varanus komodoensis* (the Komodo Dragon) is the largest surviving monitor, from three islands between Bali and Timor in the Indonesian archipelago. This 'dragon' was to become important to me.

While *Megalania* specimens are rare, we know that this monitor and *Genyornis* co-habited east-central Australia. My concept for the image of this bird and reptile did not immediately rely on any inferences about the lifestyle of the bird, although I knew there was considerable evidence that could be brought to bear on my reconstruction. However, I first turned to the behaviour of modern varanids for inspiration. Extensive studies of Komodo Dragon populations and their activities were my primary source, together with a little of my own experience with Gould's Sand Monitor and the more arboreal Lace Monitor, both from eastern Australia.

Like adult Komodo dragons, *Megalania* was too big to climb trees. These large extinct reptiles were most likely terrestrial ambush predators and as such were just as likely to be able to take large animals as the Komodo Dragon does today. The dragons are remarkably efficient at this and have developed complex behavioural strategies to subdue their unfortunate prey. Given that the largest living member of the monitor group is not an obligate scavenger, as was once commonly thought, we suspected that neither was *Megalania*. Furthermore, there are a number of specialised features of *Megalania*'s skull, along with its serrated teeth, that make the active predator hypothesis quite plausible.

Megalania lived at the time of the Megafauna and, in fact, was part of it. There had been a diverse, contemporaneous radiation of large browsing and grazing marsupials on which *Megalania* could prey or scavenge. Ground-nesting birds, and certainly their large eggs, would have made nice snacks! I knew from my own experience that monitors are marvellous egg thieves. Maybe this could be my narrative for including the *Megalania* and *Genyornis* in close proximity. I was looking to construct a scene with as much naturalism as I could muster. Central to this issue were not only the posture and behaviour of both animals, but also the need to render a scientifically realistic, reconstructed landscape. Fortunately, with both the animals and their environments being only a few tens of thousands years in the past, the validity of comparisons at all levels was to be less problematic than some of my previous reconstructions. Good analogues could be found in the modern world. The monitors were a case in point.

TOM: MEASURING *MEGALANIA*

One afternoon, I found myself on the campus of the University of California Berkeley with all my plans for work there fulfilled and nothing particularly more to do. A paper by Max Hecht, published not long before, about the fossil record of *Megalania,* came to mind. Max had noted that this top Australian Pleistocene carnivore was known from only about 20 percent of the skeleton, and he estimated its maximum length to have been between 5 and 7 metres. He also mentioned that it was essentially a giant goanna, a member of a group of lizards belonging to the genus *Varanus*, widespread in the Old World, and ranging in size from a few tens of centimetres to the Komodo Dragon, the largest of which are just over 3 metres long. Having nothing better to do that afternoon, I decided to take the opportunity to measure a few *Varanus* specimens, from the smallest to the largest individuals, available at the Museum of Comparative Zoology located on the university campus. I recorded 38 measurements of various parts of each skeleton and measured such things as the length of the femur and its width at each end, as well as the trunk and tail lengths of each individual.

Returning to Australia, I spent perhaps a week, while carrying out other duties, manually plotting selected pairs of measurements to work out how the proportions of *Varanus* species varied with their total size. For example, while the limbs of small species are relatively slender, for larger species such as the Komodo Dragon the limbs are much more stout. Likewise, the tail is shorter relative to the length of the trunk in larger species.

With this body of measurements, it was clearly possible to make a plausible estimate of the proportions of *Megalania*, using the known fossil bones of it as a guide. What I found, however, was that those known fossils were clearly from a number of individuals of different sizes. So, it was not possible to simply work out the size of the missing bones based on knowledge of the dimensions of the fossils available, in order to build a credible reconstruction of the entire skeleton merely by combining reconstructed bones with actual bones. Rather, I had to decide on a total length and then *reverse engineer* the dimensions of all the individual bones of the skeleton to come up with a plausible reconstruction. Because Max Hecht favoured a maximum length for *Megalania* of about 5–6 metres (although there was one vertebra which suggested to him the maximum length could have been as great as 7 metres), I decided to go with the more conservative estimate. And, in addition, there was a space in the National Museum of Victoria display hall that would accommodate a skeleton 5.5 metres long, so that determined the particular length chosen!

Working out what the 38 skeletal measurements would be on a 5.5 metre-long *Megalania* by extrapolating the measurements of the various species of *Varanus* that had been measured in Berkeley, I gave those figures to Brent Hall, a technician at the National Museum of Victoria. Using primarily the skeleton of a Komodo Dragon as a guide, Brent proceeded to sculpt each bone to the size I gave him, and then assembled the skeleton.

Figure 6.5 '*Megalania* & *Genyornis*' (1985): skeletal reconstruction for the Central Australian Pleistocene. Conté pastel on paper, 34.5 x 53 cm, Queen Victoria Museum and Art Gallery collection, Launceston, Australia (P. Trusler). *Megalania* reached up to 6–7 metres in length.

Not only was the beginning of this project quite casual, the subsequent steps were equally informal as well. There was never a budget for the project. Not only Brent's time but all the consumable supplies were made available as required. No one questioned the fact that the project was going ahead or made any assessment of how the restored skeleton would be integrated into the other exhibits in the McCoy Gallery. Once I discussed the idea with the preparation staff at the National Museum of Victoria and it seemed like a good one to them, it was accepted that the project was viable and it went ahead.

Brent and I described how this was done in the journal *Australian Natural History*, now sadly defunct.

To my mind, the locomotion of reptiles, such as monitors, lacked an aesthetic dimension that I understood for many other animals. I was going to need to put some effort into understanding them, if I stood any chance of bringing the skeleton back to life and producing final art that looked plausible. I turned to documentary video footage to enlighten myself on the movement and behaviour of the Komodo Dragon and other near relatives. Most birds, mammals and even fish tend to be active all the time, whether in the field or in captivity, and so one stands a reasonable chance of observing behaviours and locomotion first hand. For sketching, you need to be extra fast, and I had developed this skill in all my years of birding. Gaining numerous repeated observations, therefore, was not a major problem, and I was content to build up drawings in this way. With monitors, however, I needed to wait for some time to even catch any movement at all. I rarely got repeat observations, but these usually inert vertebrates were marvellous models for capturing the details of static body form – a case of endothermy versus still life! Photography, even with a motor drive, was not the best answer to my desire to document appropriate monitor movements. I needed to clearly understand sequential changes. The video freeze-frame button and varying speeds of forward and reverse provided the tools I needed. I sketched from video footage and made my notes with my finger on the remote to formulate the posture and stride phase for the reptile, and these sketches provided the blueprint for Craig Cleeland to build the armature and articulate the bones for the *Megalania* mount on which he and his father were toiling away.

How wonderful it was, after all this background preparation, to draw a full-sized articulated skeleton from life, precisely from the viewpoint that I required and not have to make any adjustments for light or posture.

The process involved in reconstructing *Genyornis* entailed two significant departures for me. The fossil material representing *Genyornis* was virtually complete. Articulated skeletons, and the accompanying gizzard stones that the birds had swallowed to grind their food, had been recovered from the dry sediments of Lake Callabonna in central South Australia by the Stirling and Zietz expedition back in 1893. During past times of lengthy drought, in the drying conditions of these ephemeral lake systems, individuals had become bogged in mud while attempting to quench their thirst, or perhaps simply through attempting a shallow water crossing. Their skeletons had remained entombed in the lake clays with their legs still pointing directly down into the deeper mud layers that had gripped them. No extrapolations were required!

The dune systems that traverse much of the same landscape also contain the fragmentary remains of their eggs, along with those of Pleistocene emus. So the basic material I needed to begin my reconstructions was at hand.

The behavioural comparisons that might, therefore, represent a startle-response from a tightly incubating bird – one which had finally lost its nerve when the slowly stalking reptile had risen for its fast attack – were not difficult to imagine. I could comprehend this, based on my own direct observation of various flightless bird species. After all, inquisitive and not-so-clever emus are easily startled. I shall say no more. A narrative was now in place with fully articulated skeletons centre stage.

The next phase of my task was to approximate the main muscle mass of each of the two species, lizard and bird, in accordance with the posture of their skeletons. Some of this could be estimated by reference to the bone morphology and evidence of the scarring that the ligament and muscle attachments produced on bone surfaces. The majority of muscular inference, however, had to be estimated using comparative and functional anatomy. Bone comparisons between fossil species can be helpful for this too, especially where closely-related species groups can be compared to the known structures of their modern descendants, or where there are more thoroughly-understood representatives, but the greater that biological distance the more unreliable the estimations become. Only in living animals can anatomical structures be fully known and understood functionally. This is difficult enough with muscle systems in living forms, and even more so for the extinct species studied by palaeontologists. Estimating the flesh mass and distribution in any respect is particularly slow and meticulous work and perpetually debatable for fossil forms.

That comparative distance was relatively small for *Megalania*, since their bones are so similar to those of the modern monitors such as those of the Komodo Dragon. So, my estimates from these modern monitors could be made with some confidence. Again, the video footage greatly assisted my comprehension of the textbook anatomy diagrams. Ultimately, this enabled me to achieve plausible nuances in the muscle tensions that were to show from beneath the scaled outer covering of the integument. All these were subtle matters, perhaps, but were so important to my final reconstruction if I stood any chance of communicating the sheer power and speed of this massive, bulky reptile with a static, two-dimensional image.

The evolutionary relationships of *Genyornis* were not so clear, a problem that was both compounded and disguised by the final layer of plumage. The decidedly delicate skulls of *Genyornis* were the least well-preserved part of their anatomy. The most complete specimen of the skull of this bird had deteriorated badly by the time I began my work. Salt crystallisation, transportation and display damage had, at varying times, further destroyed the already fractured skull and played a part in the inconclusiveness of its original condition and prevented further detailed examination and reassessment of the original description written by Stirling and Zietz in the early 1900s.

Figure 6.6 '*Megalania* & *Genyornis*' (1985): muscle reconstruction for the Central Australian Pleistocene. Conté pastel on paper, 34.5 x 53 cm, Queen Victoria Museum and Art Gallery collection, Launceston, Australia (P. Trusler). *Megalania* reached up to 6–7 metres in length.

This situation seriously called into question any muscle reconstruction, save for the fact that the tightly bound bodies of birds are relatively consistent, and that the research I put into reconstructing this bird could be virtually obliterated by a highly speculative feather structure. This remains a challenging and contentious point about the entire reconstruction process. The fossil record has profound gaps, and the data sets on either side of these gaps are of greatly varying magnitude in both quality and quantity. With new discoveries and the release of every new research paper, the gaps are diminished and the data improved. Should I confine my activities to that which was well understood? How did I really know when something was well understood? I wondered if I should attempt a contribution to solving a problem, or wait until it was solved? Is it an illustrator's responsibility, after all? Was I seeking a solution to a scientific methodological enigma? As an artist, I was well aware that the visual medium can be a psychologically powerful tool of trade. I also knew that any visual virtuosity that I might achieve could have lasting positive or negative effects. I was quick to recognise that the attributes in the work of artists influence the next generation, and these factors are seldom solely intellectual.

I am always willing to take the time to communicate something of this dilemma to any colleague with whom I work. 'I have sufficient skills to make garbage look good!' was a common thought in my mind. Tom once said to me, with a wry smile when I was obviously agonising over some impasse, 'One must not be afraid to be wrong. If you are, you won't get much done in science!' The point of this was not one of recklessness. On the contrary, it was one of building trust for each to individually contribute, with high levels of responsibility, rigour and candour. Hopefully, a full awareness of these tendencies is the first step to honing the accuracy, and honesty, of the resulting hypothesis, whether in art or science. And one must be willing to carefully consider the criticism or praise given by the viewer.

The external structure and appearance of both my reptile and my bird was the last phase I contemplated before painting began. I say 'contemplated' for good reason, because evidence for integumentary structures is so often lacking from the fossil record. Even when it does exist, researchers are usually interested in more scientifically significant parts of the anatomy, for evolutionary reasons, and such surface structures do not receive as thorough an analysis as an illustrator might hope for. In some instances, these have been simply noted at the outset of preparation and then discarded in order to resolve the deeper anatomy – bones and teeth in particular. This, too, is changing, but for my task no information could be found.

Broad comparative inferences were my only recourse. The monitor's skin was structurally going to be little different to the scale structure of modern varanids. The colour and pattern was another matter. I had to assess the nature of the variation across the modern monitors to decide what trends I might apply. My experience and research told me that paler colours were characteristic of ventral parts and darker colours of dorsal parts, and that more contrasting disruptive and cryptic patterns were common for juveniles and small species, and less contrasting markings for adults and larger species. Yellow pigments were common for pale regions and blacks for dark markings. General pigmentation tones tended to be pale greys and browns for open and arid habitat species, darker hues for forest and temperate species – and so the trends continued. Some variations applied for intra-specific habitat ranges. The detailed geometry of spots and/or stripes was frequently species-specific. Some species displayed other pigments, such as pinks and pale blue-greens, but these were much less common. Patterning was particularly characteristic about the head, especially on the expansive throats of many species. This all indicated to me the importance of such patterns for signalling behaviour – perhaps for dominance, sexual identity or aggression. From this suite of possible character conditions, I narrowed my selection for reconstructing a large terrestrial monitor in a semiarid environment. My final choices were subjective, yet conservative, and I was willing to tell the viewer why I chose what I did. It was not arbitrary – no pink polka dots.

Without any contemporary cousins to speak of at that time, the *Genyornis* was a total mystery. The plumage of large living flightless birds tends to be relatively plain or made up of various, fine cryptic patterns comprising both pale or dark spots and striations. Bird groups have spectacular exceptions. Sexual dimorphism can be pronounced, as in the case of the ostrich, where males are formally attired in black with white collar, white fluffy wings and a creamy feather-duster tail. The females have a much plainer dress code. The age classes of birds can also be so distinct, and a species can pass through several plumage types. Reclusive cassowaries have a glossy black, rain-proof thatch for adult plumage, while the precocial chicks sport bold black and white, longitudinally striped down, and orange heads. The immature birds are plain rufous brown. There can be seasonal changes, moult changes, display plumes and soft part colourations that need to be considered. In the case of cassowaries, the adults have brilliant crimson or orange and cobalt blue skin on the head and neck, and long, pendulous throat wattles to match. No fashion statement seems inappropriate, and so what was I to do with regard to *Genyornis*?

There was, however, one important biological consideration that I could apply. When birds evolve to flightlessness, their feathers relatively quickly lose their structure. The aerodynamic, stiff yet flexible contour of the feather vanes is no longer required, and so thermoregulation and colouration resume their primary function. The plumage becomes loose, and the feather shafts more flexible. The general appearance becomes either more fluffy or filiform. Another matter related to this is that large birds, in particular, often have sparse plumage about the head and neck, and may show large areas of naked skin. While this might often be of importance in display, it might have a dual purpose in the larger ground birds – one of cooling. And this could be a real physiological necessity for *Genyornis*, inhabiting some of the late Pleistocene environments that it did.

The habitat of the scene was built up from palynological and palaeoclimatic data at hand. The sediments at Lake Callabonna that had entombed *Genyornis* contained layers of information about the floristic structure of the habitat, because they contained persistent pollen and spores that were borne on the winds and carried in by the flood-waters of the time. The fossil remains of the plant groups, and in some cases the same

Figure 6.7. '*Megalania* & *Genyornis*' (1985): integument and scenic design for the Central Australian Pleistocene. Conté pastel on paper, 34.5 x 53 cm, Queen Victoria Museum and Art Gallery collection, Launceston, Australia (P. Trusler). *Megalania* reached up to 6–7 metres in length.

(Previous page) **Figure 6.8** '*Megalania* & *Genyornis*' (1985): reconstructed scene of the Central Australian Pleistocene. Watercolour and gouache on paper, 34.5 x 53 cm, Queen Victoria Museum and Art Gallery collection, Launceston, Australia (P. Trusler). Final art of *Megalania* and *Genyornis*.

modern species preserved in these sediments, could be identified. The climatic data recorded by interpretation of the mineralogy and sediments indicated the degree of aridity that prevailed. The known habitat preferences of key species of plants and plant communities, recognised in modern floras, could be compared with the fossil samples to attempt reconstruction of the ecological setting occupied by *Genyornis*.

After all this background research, the painting could finally begin, and the proposed stage presentation would showcase each step of the process, evolving from one to the next to finally reveal the fully-coloured watercolour image. The style chosen for this was to be photorealistic, in an effort to convey some sense of immediacy, or freeze of the fast action. An impressionistic or more painterly style would not carry the same illusion. While I could include imagery of plants, such as *Pittosporum, Callitris* pines and *Allocasuarina* in the background with salt bush and desert everlastings in the foreground, I felt a narrow plane of focus would be better employed to suggest the vegetation floristics and structure without the need for excessive detail. This would better direct the viewer's attention to the protagonists, an efficiency that I hoped would strengthen the narrative without trivialising the information.

There is an important footnote I can add to my comments here. In addition to Tom's input on *Megalania*, I had been seeking advice from Ralph Molnar of the Queensland Museum, Brisbane. Shortly after he had critically reviewed my preliminary work, I received a package from Ralph with a letter filled with excitement. He had just unearthed some new skull elements of *Megalania*. One of these demonstrated that, unlike any other monitor known, there was a bony vertical projection in the centre of *Megalania*'s forehead. *Megalania* had a crest that no one could have predicted! The cast in the package demonstrated that the original bone had the faint indentations of overlying scales. Ralph's notes, the basis for his ensuing paper, placed the osteology in clear context, and so I corrected my drawings, and Craig took up sculpture in his mounted skeleton. This new evidence at last gave me a small glimpse of integument and radically altered the profile of my *Megalania*'s head. This crest did not likely have a primarily structural function; rather, it was probably a behavioural feature. This reinforced my predictions for depicting contrasting pattern on the head. Perhaps bobbing or nodding was an important part of *Megalania* behaviour?

In my reconstruction, *Megalania* turned out to be much more impressive than I could have hoped. But the story doesn't end there. What of the hapless bird? This was something that was going to come to light many years later (*see* Chapter 9).

For a realist painter, one who tends to dwell in the realm of external surfaces, the issue of inventing them was novel. The palaeontological reality was almost exclusively internal, but subsequent projects (e.g. reconstructions of the Ediacarans – *see* Chapter 11) were to see that change, in intriguing ways. The conclusion of the decade for Pat and Tom saw them feverishly working on *Wildlife*. It was Pat who requested permission to use *Megalania* and *Genyornis* for cover art and proposed two new, additional reconstructed scenes. Would I be interested if she could secure the funds from the publishers? She argued the case successfully and presented me with the brief for works to depict aspects of the Paleozoic and Mesozoic from an Australian perspective. The book was conceived to present a lavish photographic survey of the Gondwanan palaeontological fauna, as the illustrative complement to an evolution text. Realistic artwork was, therefore, stylistically desirable to contribute hypothetical glimpses, measured and scientifically based, into three evolutionary eras. My first reconstruction of *Megalania* and *Genyornis* was eminently suited to represent the Cainozoic.

The renowned Gogo site from northwestern Australia (an ancient geological reef that was later uplifted to form the existing mountain range on the pastoral property of Gogo Station) was destined to represent the Paleozoic. A steady stream of exquisitely preserved fish fossils was being collected from this locality, and their careful acid preparation was producing revelations about evolutionary trends during Late Devonian times. The preservation mechanism with nearly complete fossils encased in limestone nodules has produced uncrushed fossils preserved in three dimensions. Armoured integuments were common for the fish groups of the day – a biological equivalent to the medieval military strategy of protective plates and chain-mail. All of these had been preserved in exquisite detail. What a delight! I would be dealing mostly with external surfaces for the first time.

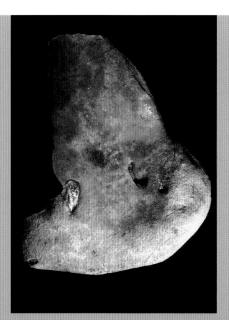
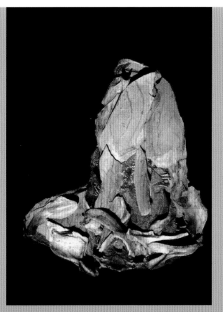
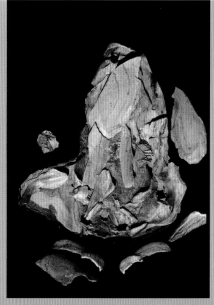

Figure 6.9 Gogo nodule under acid preparation. Slow etching with dilute acetic acid gradually exposes the bones of the ancient fishes that lived in the shallow marine reef environment of now Western Australia. The width of the nodule is about 32 cm.

The two species at Gogo chosen to feature in the Devonian scene were the well studied *Griphognathus whitei* and *Eastmanosteus calliaspis*. They were also well known from the USA and Scotland, and their cosmopolitan distribution provided an important opportunity to fuse the information available from specimens preserved in different ways. Firstly, I had to learn much more about fish than my formative science background had provided, and so Pat and Tom sent me off to discuss this with John Long at the Western Australian Museum, Perth, Alex Ritchie at the Australian Museum, Sydney, and Prof. Ken Campbell and his team at the Australian National University in Canberra. They willingly obliged, and pushed me along another steep learning curve. Of course, I needed to see their institutions' collections for myself and get to know the tangible remains of the fishy suspects.

My research would also take me to aquaria and sanctuaries to observe fish behaviour and wonderful 'living fossils' like *Neoceratodus*, the Queensland lungfish. I also sketched the famous *Latimeria* from video footage to enhance my understanding of locomotion in another lobe-finned group, the coelocanths. *Latimeria* is the single living relative of this once widespread group. Aside from needing to understand the biology of the fish that might be useful for comparative purposes, the reconstruction could proceed with a little less emphasis on an overall anatomical assembly. There were telling internal features that provided sound scientific clues to a wide variety of aspects about the fish in question, and the environment in which they lived and died was well understood.

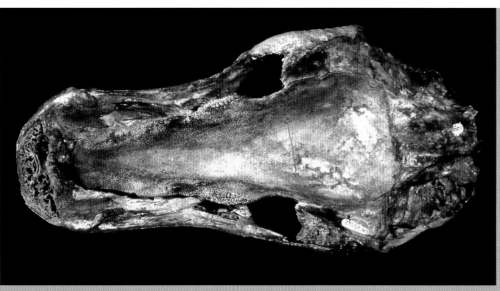

Figure 6.10 The Devonian lungfish *Griphognathus whitei* from the ancient reef found on Gogo Station, Western Australia, the basis for part of Peter's reconstruction of this ancient marine environment that existed more than 360 million years ago. The length of the skull is about 16 cm (photo F. Coffa, courtesy of K. Campbell).

For example, when I first envisaged a Devonian reef scene, images of the Great Barrier Reef sprang to mind – a place infused with colour, light and a spectacular mosaic of refracted optical effects penetrating a shallow tropical blue sea. The data coming from the fossils was to abruptly shatter that illusion.

The fossilising process had occurred in the calcium carbonate–rich sediments deposited on the floor between reef platforms. This ancient seabed topography is perfectly visible at Gogo today. The lungfish, *Griphognathus*, clearly did not restrict its activity to the shallows. The detail of the well-vascularised anatomy of its gill apparatus revealed that this lungfish was not an air gulper, unlike the few surviving members of this group. *Griphognathus* also had a highly sensitive duck-bill that must have been functionally akin to the platypus snout. The array of nerve receptors in the *Griphognathus* rostrum were preserved in microscopic detail. This fish had specialised grinding tooth plates and cheek pouches to process invertebrates or shellfish, which could easily have been shovelled from beneath the sediment. All the attributes of this lungfish pointed to a bottom-dwelling lifestyle. When I factored into my thinking the palaeo-reef topography, this sank my viewpoint for the scene to depths where the light levels were low. It also became obvious to me that a deeper setting was a key aspect to be communicated about the evolutionary story coming from this site, and that the gloom in which the artist had now found himself was good!

The other fish of choice was *Eastmanosteus*, a mid-water predator. It was a formidable looking fish. Its head and trunk were packaged in finely-textured dermal armour plates, with apertures for large eyes, shark-like pectoral fins and serious-looking pincer jaws. The inclusion of *Eastmanosteus* did not need to alter the context of the setting, however.

The environmental data and contemporaneous invertebrate fauna from the site were examined. Ken Campbell noted that some light must have been reaching these depths, evidenced to him by what he thought were the fossilised remains of algal mats on the sea-bed. This also meant the water clarity was probably better for my purposes too – avoiding the need to invent a 360 million year old flash-light for my realism! Some compromise was ultimately needed. I raised the predicted light levels slightly in order to ensure that the Gogo denizens could be seen clearly. There was one advantage in wanting to depict the depths in natural low-light levels. The light wavelength spectrum is filtered to blue, and so at depth the world becomes monotone, and I did not have to interpret colours.

Neil Archbold guided me to references illustrating the beautiful types of geometric patterns that were possible on the Paleozoic cephalopod shells he knew so well. Again, these things are rarely preserved in the fossil record, and the Gogo taphonomy recorded no external pigmentation. I could take some clues for *Griphognathus* from the modern lungfish, but there were no such possibilities for placoderm fish like *Eastmanosteus*. Unpreserved details like the fin shapes could be ascertained from the flatter preservation of this form known from other sites. I could proceed with the painting and chose to do this in oils in order to best capture the intensity of the deep blue. I picked up my brush and began.

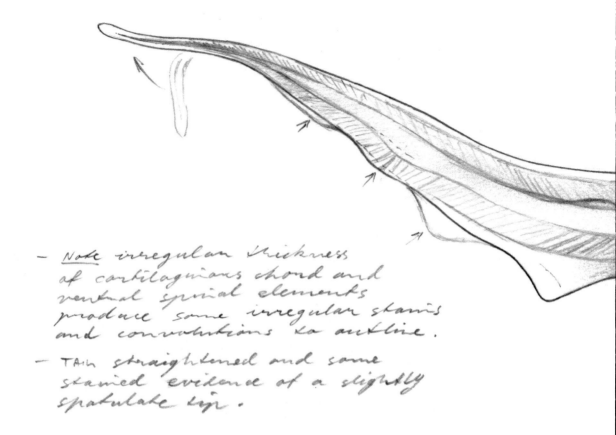

Figure 6.11 *'Coccosteus – study notes'* (1990): Australian Museum, Sydney. Graphite on paper, 28 x 38 cm, collection of the artist (P. Trusler). A comparative study of the placoderm, *Coccosteus cuspidatus*, from a two-dimensional mode of preservation that shows traces of the fin outlines of these Devonian fishes.

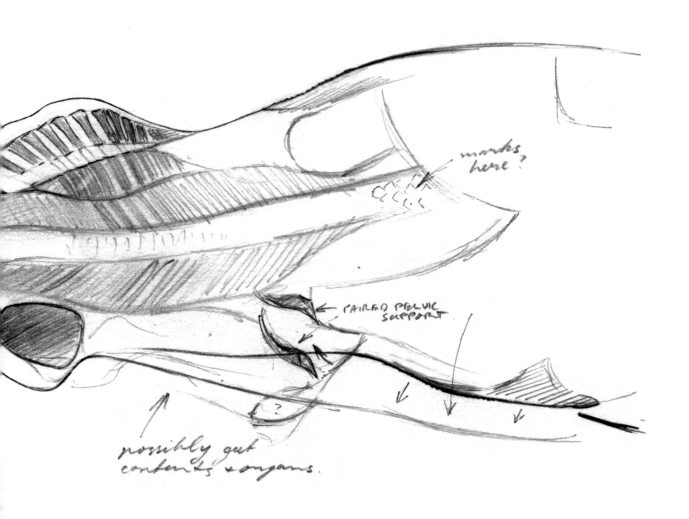

Figure 6.12 'Gogo reef reconstruction' (1991). Alkyd oil on linen over hardboard, 51 x 71 cm, private collection (P. Trusler). The disturbance of the feeding behaviour of *Griphognathus whitei* has attracted the attention of the predatory arthrodire placoderm, *Eastmanosteus calliaspis* and a variety of invertebrates such as nautiloids. The shells of bivalved arthropods and a nautiloid can be seen on the sea bed. The length of the lungfish skull is about 16 cm.

7
The Dinosaurs of Darkness

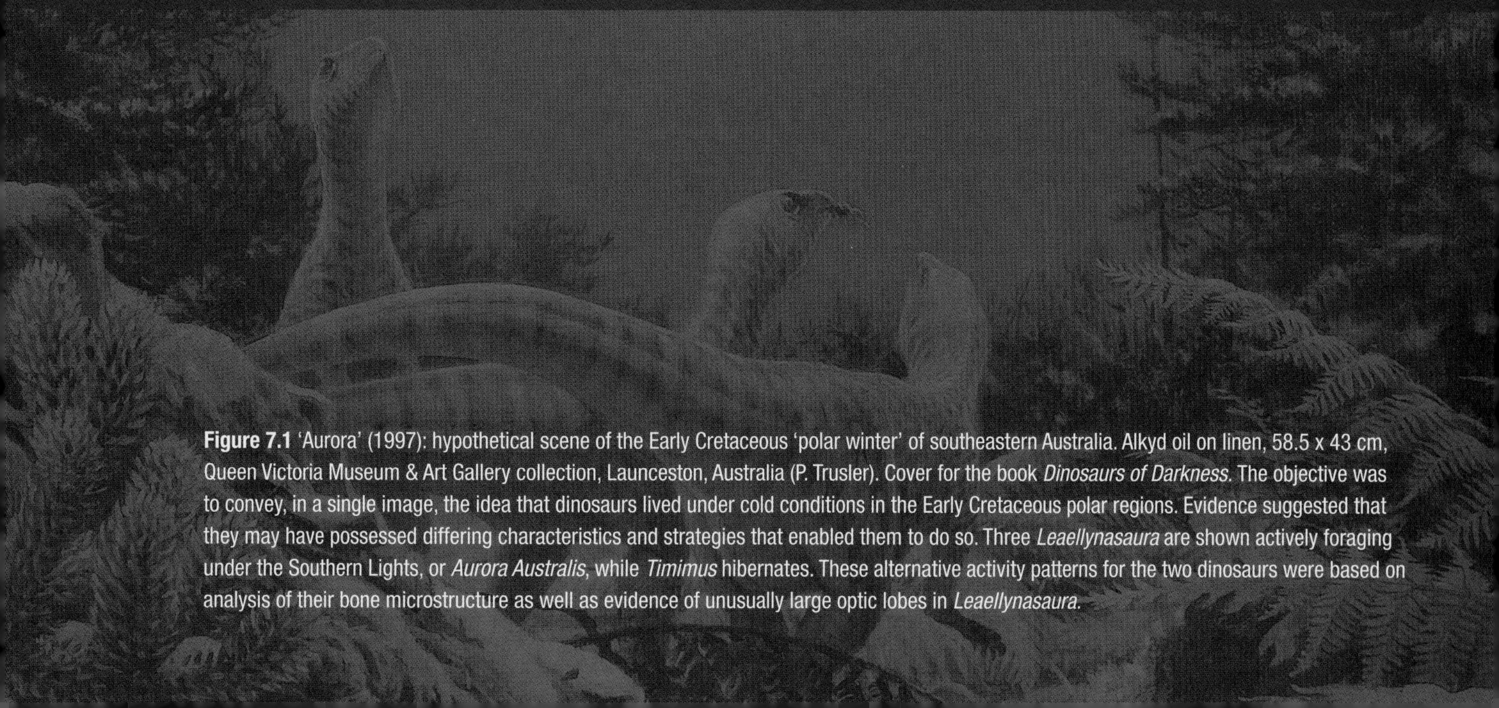

Figure 7.1 'Aurora' (1997): hypothetical scene of the Early Cretaceous 'polar winter' of southeastern Australia. Alkyd oil on linen, 58.5 x 43 cm, Queen Victoria Museum & Art Gallery collection, Launceston, Australia (P. Trusler). Cover for the book *Dinosaurs of Darkness*. The objective was to convey, in a single image, the idea that dinosaurs lived under cold conditions in the Early Cretaceous polar regions. Evidence suggested that they may have possessed differing characteristics and strategies that enabled them to do so. Three *Leaellynasaura* are shown actively foraging under the Southern Lights, or *Aurora Australis*, while *Timimus* hibernates. These alternative activity patterns for the two dinosaurs were based on analysis of their bone microstructure as well as evidence of unusually large optic lobes in *Leaellynasaura*.

The Scientist: Tom

On a cold, blustery day in November 1978, the geologist Rob Glenie was giving me some respite from the overwhelming enthusiasm of two cousins – John Long and Tim Flannery – by taking the two youngsters on a trip to the site where the geologist William Hamilton Ferguson had found Australia's first dinosaur specimen on 7 May 1903. That fossil, a partial toe bone of a carnivorous theropod, remained the sole dinosaur to have been identified from Victoria for the next three-quarters of a century. Almost as soon as they reached the shore platform near a rock stack called Eagle's Nest, a few kilometres from the town of Inverloch, John found a pebble with a bone fragment in it. Over the next six months his cousin Tim continued to prospect all the Cretaceous outcrops between Eagles Nest and the town of San Remo. He recovered about 30 fossil bones, including an ankle bone (astragalus) of a carnivorous dinosaur, plus a distinctive femur of a herbivorous one.

Tim also turned up a most enigmatic fossil fragment, which would not be confidently identified for a decade; in the meantime, we informally referred to it as the GOK ('God only knows'). In the end, after many more specimens of the animal had been found, it became evident that this fossil belonged to a group of amphibians, called temnospondyls. In 1978, when Tim found it, this group was thought to have become extinct 80 million years before the sediments that enclosed the GOK were deposited. Although Ralph Molnar and Pat had independently suggested it might be such an amphibian, the confirmation that this was the case came from the numerous discoveries of additional specimens by Mike Cleeland, one of our volunteer prospectors. These had unequivocal features to be found only in temnospondyls. The scientific name finally applied to these fossils, in part to honour Mike, was *Koolasuchus cleelandi*. The generic name *Koolasuchus* can be translated as 'cool crocodile' – crocodile being the closest word in either Greek or Latin that could be applied to these amphibians. The generic name in this case honours Lesley Kool, who prepared these specimens and the majority of the Cretaceous fossils collected by our team over nearly three decades. The coincidence of her surname and the frigid climatic setting where this temnospondyl lived was a nomenclatorial opportunity that simply could not be missed!

To understand *Koolasuchus cleelandi* as a living animal, Pat, Peter, our daughter Leaellyn and I travelled to the Asa Zoo in Hiroshima, Japan. There, the staff took great pride in introducing us to the living giant salamander native to the area, as well as its vicious Chinese relative. These are the closest known living analogues to *K. cleelandi*. Not only were we able to observe them within the confines of the zoo itself but, both during the day and at night, the zoo staff kindly took us to observe these animals 'in the wild' – i.e. the rice paddies and closely-settled hills near the city.

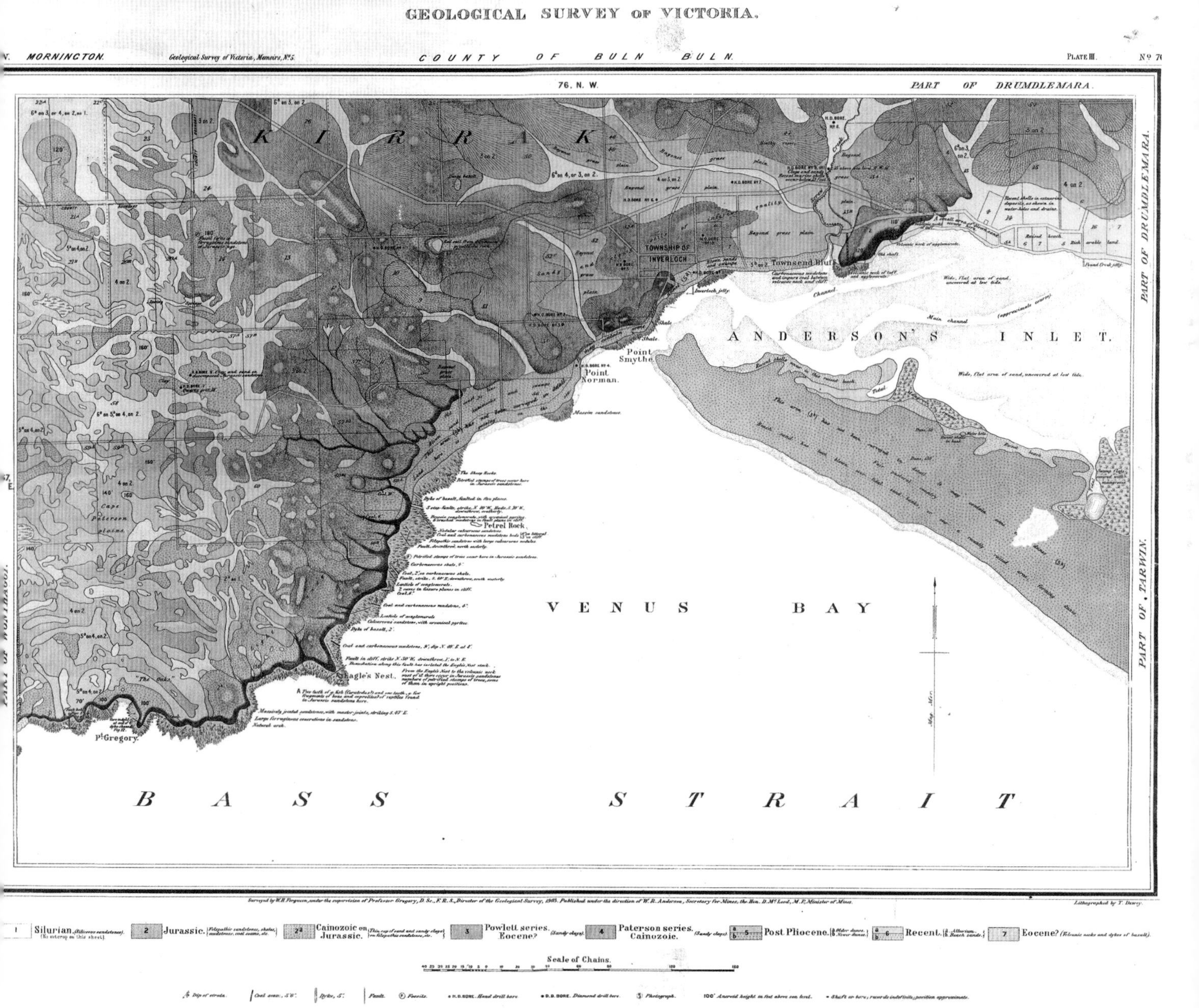

Figure 7.2. Hand coloured geologic map of the area around Eagle's Nest near Inverloch prepared by William Hamilton Ferguson, who found the first dinosaur in Australia on 7 May 1903. Ferguson was compiling these maps for the Geological Survey of Victoria, part of a program to record the occurrence of geological resources in the state.

Koolasuchus was the first of a number of revelations about the nature of the land vertebrates occurring in the Cretaceous rocks of Victoria that were to surprise us over the years.

As the study continued of these amphibian fossils, together with the dinosaurs and other vertebrates that lived alongside them, we soon realised that they had another interesting aspect. When the animals whose fossil bones we found were living and dying, their habitat lay within the Antarctic Circle. Somewhat later physical evidence would be found to suggest that southeastern Australia was one of the coldest places on Earth during the time dinosaurs were alive. Nowhere on Earth except in southern Victoria have sediments of this age been found that reflect the former presence of permafrost. However, there is other evidence of extremely cold conditions in this region. Further from the South Pole of the day, rock sequences of the same age in northern South Australia contain drop stones. These seemingly out-of-place boulders embedded in fine-grained marine shales were deposited on the floor of a shallow sea. Evidently, they fell from above when glacial ice floating on the surface of the ocean in which these rocks were embedded melted, releasing the boulders. Those same sea-bed deposits contain the mineral calcite in a shape that indicates it was originally deposited as the mineral ilkaite. Ilkaite only forms in waters where temperatures are close to 0°C and changes to calcite as waters become warmer, retaining the crystalline shape of its predecessor.

Figure 7.3 Rock structures near Inverloch, southern Victoria. These sedimentary structures indicate that when this rock was unconsolidated sands and muds, it was continuously frozen for more than two years. That is, it was permafrost. It is the only known evidence for the presence of permafrost anywhere on Earth during the Mesozoic era, the time when dinosaurs lived (photo by A. Constantine).

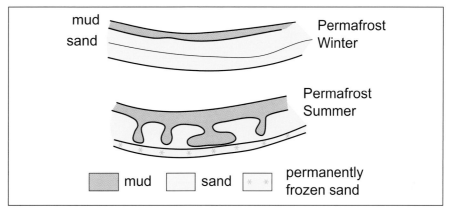

Figure 7.4 Diagrams showing how cryoturbation structures are formed. The formation of a drop soil structure in seasonally frozen ground below a shallow pool. The top diagram shows a cross-section of a frozen shallow pool. The top layer is a compact clay laid down at the bottom of the pool. Beneath it are two layers of frozen sand, perhaps as much as 50 % water by volume. In the lower diagram, the upper layer of frozen sand has melted to become a slurry in the summertime. The clay above, having less water in it than the slurry, is denser and, therefore, tends to sink into the sand-water slurry until it reaches the still frozen, deeper layer of water-saturated sand. When the clay reaches the still frozen sand, it can no longer sink any deeper and tends to flatten out against its upper surface. (From *Dinosaurs of Darkness* by T.H. Rich and P. Vickers-Rich.)

Such indicators of cold conditions during the Early Cretaceous of southeastern Australia reinforce the palaeomagnetic data, gathered by geologists using a variety of procedures, which point to a palaeolatitude between 70°S and 80°S. Their estimates were based primarily on synthesising data from several continents, assuming a particular fit between Australia and Antarctica. To further check these hypotheses, Michael Whitelaw – a one-time student of Pat's working in Neil Opdyke's laboratory at the University of Florida – analysed more than 600 samples from the very rocks that produced the Victorian dinosaurs. His work supports this high palaeolatitude estimate.

This polar habitat has spurred on investigations of these fossils for the past three decades – a high-latitude, often frigid environment at a time when dinosaurs thrived on a globe generally experiencing 'greenhouse' conditions!

After Tim Flannery's initial prospecting of the coastal outcrops between Inverloch and San Remo, I realised that to the west of Melbourne, in the Otway Ranges, there were more coastal outcrops of Early Cretaceous rocks. So, in 1979, together with a group of colleagues, I began a systematic survey of these rock sequences. The following year, we reached a then unnamed cove. Tim Flannery and Michael Archer were walking together at the base of a cliff while I was searching outcrops nearby at the sea's edge. Suddenly, they let out a whoop, dropped to their knees, and in a short time had found four amorphous bone fragments each about the size of an adult human thumbnail. Needing a name for this unnamed cove to refer to it in my field notes, four nights later, I casually typed in 'Dinosaur Cove', never thinking that it would assume the significant role it did in Pat's and my scientific lives. It would be more than three years before a definite dinosaur bone would be found there. But in the meantime, in the confidence that such fossils could be there, the name 'Dinosaur Cove' was proposed to the Place Names Committee of the Victorian government and officially accepted. We had put the place on the map! My optimism was justified in 1984.

It was the last place where I wanted to excavate fossils. I hoped to find another site that would be much less difficult to excavate and definitely more accessible than Dinosaur Cove. From the outset, I realised I would have to tunnel to effectively work this site. I knew absolutely nothing about tunnelling and blasting, and had no desire to learn anything about these techniques through first-hand experience. However, I had not factored in the enthusiasm of the general public in charting the course of my own research and fieldwork.

It just so happened that the Friends of the National Museum group was organised at about the same time as Dinosaur Cove was discovered. During the *Dinosaurs from China* exhibition, the members of this group served as volunteer guides, and also set up a shop selling dinosaur paraphernalia. With the money they made, they were keen to go on a dinosaur dig. I tried to dissuade them, telling them that there was no place in Victoria where this was practicable. However, they knew that through the work of Pat, Tim, Mike Archer and others, dinosaur bones had been found during the previous few years in both the Otway and Strzelecki ranges and they would not be put off.

That led, in February 1984, to the 16 maddest days of my life. With 65 amateur volunteers aged from seven to 70, an attempt was made to tunnel at Dinosaur Cove for the purpose of exposing a layer of rock that was once soft sand at the bottom of an ancient stream channel. Rock drills and breakers were supplied by Atlas Copco, a Swedish mining equipment company, whose staff had never dealt with a group of people like us: volunteers with only the slightest idea of how to properly use the gear. Fortunately, a few among the volunteers were able to quickly work out how to operate the compressors, rock drills, and rock breakers. Despite many breakdowns owing to human inexperience and the unsuitability of some of the equipment in the particular circumstances at Dinosaur Cove, and by working 24 hours a day when possible, we slowly excavated an adit about the size of two telephone booths out of the hard sandstone. By means of all this work, we recovered 85 fossil bones and bone fragments, which included the first undoubted dinosaurs from Dinosaur Cove. It was then clear that with a concerted effort, more dinosaur bones and teeth could be systematically recovered from the site. Because I now had this hard-won piece of precious knowledge that there were actually bones in place there, I was unable to turn my back on Dinosaur Cove.

Every summer bar one for the next decade, there was a dig at Dinosaur Cove. Over the years, the excavation technique improved, and became more efficient. More elaborate techniques and equipment were employed, one being a flying fox. This was an aerial tramway conceived, designed and constructed by one of the volunteers, John Herman. With the 'fox' in place, it relieved us from the necessity of carrying the heavy tunnelling equipment on our backs down a 90-metre cliff (nearly 300 feet). In 1987, we began blasting to remove overburden above the fossil layer, making it possible to cut deeper tunnels. Volunteers were the mainstay of this project. They contributed more than 20 person-years of hard labour to the work at Dinosaur Cove, and all their work was unpaid.

As the work progressed, we realised that the fossil bones were concentrated in the first 10 centimetres of sand immediately overlying a thick claystone. The claystone represented floodplain sediments laid down far away from an active channel, at a time when the water in a pre-existing stream overtopped its banks during a flood. Subsequently, a new channel was cut across the area where the modern fossil site was located. The sand, mixed with chunks of clay ripped up from the flood plain along with bits of carbonised wood and the occasional bone, was deposited in this new channel. We were now, more than 100 million years later, digging into those soft muds and sands, which had turned to hard rock as the original sediments were deeply buried. Only when the Otway Ranges subsequently rose above sea level were those ancient sediments exposed. Bones preserved in these rocks had been transported to their final resting place as individual fragments, not as whole skeletons.

Above the base of the fossil-rich channel, the sediments fined upward, and debris content declined. These finer-grained sediments were laid down where the current had diminished greatly and, instead of being in an actively flowing stream, they accumulated in a quiet 'billabong', or oxbow lake, cut off from the main stream. It was in the oxbow environment that two partial skeletons of the dinosaur later named *Leaellynasaura amicagraphica* were found. The first specimen was collected in 1987.

Figure 7.5 The skeleton of the dinosaur *Leaellynasaura amicagraphica*. It was named after Pat and Tom's daughter Leaellyn, who wanted her own dinosaur, and also the Friends of the National Museum of Victoria (*amica*) and the National Geographic Society (*graphica*), both of which groups were instrumental in the finding of this skeleton, one of the few to have been found in Australia. The femur is about 13 cm long (photo by S. Morton).

(Previous page) **Figure 7.6** '*Leaellynasaura amicagraphica*; spring river scene' (1992): reconstructed scene of the Otway Basin, Early Cretaceous of Victoria, Australia. Alkyd oil on linen, 30.5 x 43 cm, National Geographic Society collection, Washington (P. Trusler). Two individuals of *L. amicagraphica* on a sandbar near a flood-swollen river 106 million years ago in Victoria. Large rivers were a prominent feature of the flood plain formed in the rift valley between Australia and Antarctica in which the dinosaurs known from Victoria lived. The dinosaurs would have been about 1 metre in length.

The bones found in the sediments below the skeleton-bearing fine-grained sands of the oxbow had been preserved differently, having been transported some distance in rapidly flowing water. Unlike the disarticulated bones, the two skeletons that were given the name *L. amicagraphica* died close to, if not in, the oxbow where the fossil remains were found 106 million years later.

In reconstructing an image of the first known skeleton of *L. amicagraphica*, shown in the process of becoming a fossil, Peter portrayed the corpse lying on the edge of an oxbow lake about five metres wide, reflecting much of the regional geological information concerning the environment in which she lived. Around her in the reconstruction, Peter painted debris of the sort we knew to have been around in the streams when she was alive. In order to represent the plants that were her contemporaries, which were decidedly larger than the young *L. amicagraphica* specimens, Peter portrayed the flora as reflections in the water of the lake. In this way he was able to combine images of objects of much different size without sacrificing

Figure 7.7 '*Leaellynasaura amicagraphica* – corpse' (1990): skeletal reconstruction for the Otway Basin, Early Cretaceous of Victoria, Australia. Graphite on paper, 56 x 66 cm, private collection (P. Trusler). The reconstruction drawing prepared to estimate the proportions of the entire skeleton: the pose is typical of the post-mortem tension imposed by the contraction of muscles and ligaments during rigor mortis. Total length is about 1 metre.

Figure 7.8 Leaellyn Suzanne Vickers Rich holding the skull of the little dinosaur that bears her name. Leaellyn accompanied her parents on many expeditions from an early age (3 months) and herself became a palaeontologist as a hobby, though her profession is law (photo by P. Menzell).

Figure 7.9 Skull of *Leaellynasaura amicagraphica*, showing the internal impression of the brain. Found at Dinosaur Cove, associated with a partial skeleton in an isolated billabong that contained no other fossil remains except this little dinosaur. The water in this billabong dried up more than 106 million years ago, encasing her bones, which were gradually buried about two kilometres underground, before being forced up into the Otway Ranges over the last few million years, then exposed and found by the diggers at Dinosaur Cove. The length of the skull is 52 mm (photo by F. Coffa).

details of the smaller one that was the primary focus of the painting. Because there was no indication of what time of year *L. amicagraphica* actually died, we made an arbitrary choice to place this corpse in an autumn setting. This was done so that a thin layer of ice indicative of a cold environment could be plausibly placed on the water. The light in the sky was given a quality expected at this high latitude at that time of year. Another unknown factor in this reconstruction was the colour pattern and skin texture on the little dinosaur. As fossil skin impressions are known for hadrosaurs, or duck-billed dinosaurs – reasonably close relatives of *L. amicagraphica* – those were used as guides for the texture. The colour pattern was based on that of modern reptiles inhabiting forested environments somewhat akin to the gymnosperm-dominated forests of the Early Cretaceous of southern Australia. *L. amicagraphica* was Peter's first attempt to reconstruct and illustrate a dinosaur as a complete animal. He deliberately chose a corpse because the issues of the pose of the animal were easier to resolve in this, his initial illustration of this nature. As closely as we could determine, it was set in the place in the ancient oxbow lake that the first specimen of *L. amicagraphica* was buried. In the illustrations to follow, Peter built on this experience, and subsequent restorations were of active, living animals.

As tunnelling continued at Dinosaur Cove, we exposed about 100 square metres of tunnel floor and found several new dinosaurs. However, mammals and birds, which presumably lived alongside these dinosaurs, continued to elude us. We knew that during the Early Cretaceous, birds, mammals, and non-avian dinosaurs occurred together elsewhere in the world. Although we were very pleased to have these dinosaurs, we never lost sight of our original goal – to find mammals and birds.

Figure 7.10 Excavation and processing of sediments of the Cretaceous age at Flat Rocks in the Strzeleckis (Gerrit Kool, centre), and at Dinosaur Cove in the Otway Ranges of Victoria (Natalie Schroeder jackhammering, left), laid down more than 105 million years ago in a great rift valley that separated Antarctica from Australia. (Right) Lesley Kool preparing fossils (photos by F. Coffa and L. Kool).

Work at Dinosaur Cove ceased after 1994, because the only remaining fossil-bearing rock that we knew of was at the bottom of a hole, 5 metres below sea level. The fossiliferous rock layer dipped downward towards the centre of the cove, and in the final stages of the work three pumps working simultaneously could barely keep the seawater to a manageable level.

Two years after all work ceased at Dinosaur Cove, Lesley Kool recognised a fossil mammal from there in the collection held temporarily at Monash, where she was working. It was a monotreme, represented by a humerus closer in appearance to that of an echidna rather than a platypus, and yet not clearly a member of either family. It was distinct enough to rate its own name, *Kryoryctes cadburyi*, 'Cadbury's cold digger'. The species name recognised the assistance of the chocolate manufacturer Cadbury, who provided the discoverers of this first 'polar' monotreme with a cubic metre of chocolate – thereby honouring a promise that I had foolishly made long before the discovery. This first mammal fragment was interesting, but not the radically different kind of mammal I thought must have been present in Australia during the Mesozoic.

Only a few kilometres northeast of where Ferguson had found the first Australian dinosaur in 1903, another fossil site, at a place called Flat Rocks, was discovered in 1991 by two of the stalwarts among the volunteers from Dinosaur Cove, Lesley Kool and Michael Cleeland. At last, this was the kind of site I had hoped to find before I was dragooned into 'mining' at Dinosaur Cove. This site was in the open – no tunnelling was necessary! – and access was so much easier too. But if it were not for the hard work at Dinosaur Cove, I would likely never have met the two people who made the discovery of this site near Inverloch. Likewise, they probably never would have had any involvement in the search for the dinosaurs and other animals that lived alongside them in Victoria. I have often thought that the most important discoveries at Dinosaur Cove were not dinosaurs at all, but highly motivated people.

Work commenced at this new locality in 1992. The diverse dinosaur fauna found there complemented that from Dinosaur Cove. The fossils from Flat Rocks were 10 million years older than those from Dinosaur Cove, so comparisons of those two fossil assemblages gave new insights into how a polar dinosaur community changed over a significant period of geological time, as temperatures rose.

Work at Flat Rocks continues to this day – largely because on 8 March 1997 a sharp-eyed volunteer, Nicola Sanderson (*née* Barton), cracked open a walnut-sized piece of rock and spotted a few teeth in a broken bit of jaw. The tiny teeth, having two roots instead of one, were unlike any dinosaur of that size. In fact, they belonged to a mammal! But, unlike the humerus from Dinosaur Cove recognised the year before, these teeth did not belong to a monotreme. Being the hardest fossil to find – the first one – Nicola's discovery paved the way for systematically recovering more of the long-sought Australian Mesozoic mammals. Over the years, upwards of 46 mammal jaws have been recovered from the Flat Rocks site. Their identity has challenged prevailing ideas about mammalian evolution on the Australian continent over the past 110 million years (*see* Chapter 12). Instead of being the expected marsupials or monotremes, they were predominantly an entirely new group of placentals, or possibly even representatives of a group of mammals of equal rank and never before seen.

118 THE ARTIST AND THE SCIENTISTS

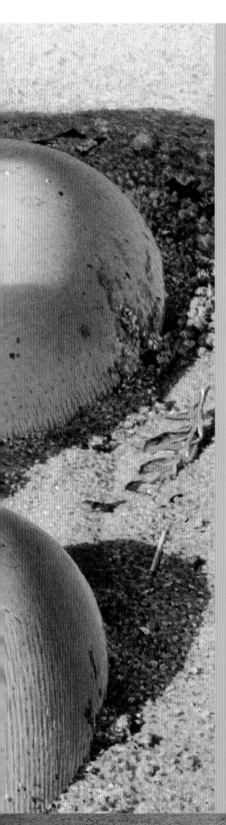

Figure 7.11 *'Leaellynasaura* hatchling' (1993): product illustration for *Australia's Dinosaur Era* philatelic issue. Watercolour and gouache on paper, 20 x 32 cm, Australia Post collection (P. Trusler). Peter's supplementary illustration for the stamp release featured on the cover of the various southwest Pacific editions of *Time* magazine on 9 October 1993, at the same time as the movie *Jurassic Park* opened and the *Dinosaurs of Australia* postage stamps were launched. It was a bumper year for his art, and for our scientific research outcomes.

With the release of *Jurassic Park* in 1993, that year was definitely the 'Year of the Dinosaur'. Australia Post approached Pat about producing a set of stamps about Australian dinosaurs; the southwest Pacific editions of *Time* magazine ran two feature articles about 'Dinosaurs downunder'; and Museum Victoria hosted a dinosaur exhibition, *Great Russian Dinosaurs*. All of this led to a flurry of further work for Peter.

In the production of all of these reconstructions, there was a constant interaction between the three of us, as we first discussed the objective and then how it was to be visualised. Part of this process involved Peter drawing preliminary sketches, which Pat and I then commented on, and he in turn responded with pencil and brush. In this way, Peter's ideas gradually took the final forms expressed in the finished artworks.

For the Australian dinosaur stamp issue a new requirement had to be met – the needs of Australia Post. Their staff wished this major issue to be an Australian dinosaur stamp issue, not just Victorian dinosaurs. So, included in the principal image was a restoration of *Muttaburrasaurus langdoni,* a dinosaur from Queensland that was one of the only two Australian dinosaurs then known in their entirety. All the remaining dinosaurs in the stamp image were known from Victoria, although in one case a few isolated bones of an armoured dinosaur that had been recovered from Victoria seemed related to a Queensland form, *Minmi paravertebra*, known from a few skeletons, so *Minmi*, too, was included. The basis for the carnivorous dinosaur, now thought to be a distant relative of *Allosaurus*, was the single astragalus, or ankle bone, found years earlier near Inverloch by Tim Flannery. So, the entire skeleton of the well-known North American *Allosaurus* had to provide the basis for its appearance in the stamp issue. Similarly, representation of the dinosaur *Timimus hermani* – an ornithomimosaur, known from a few limb bones – was based primarily on the skeleton of the Mongolian *Gallimimus bullatus*. This was done for the simple reason that a mounted cast of *Gallimimus* in a Museum Victoria display had a femur similar to that of *T. hermani*.

As these dinosaurs had to be placed in an environmental setting, we used all the geological data at hand to set them in a Victorian Cretaceous scene. The particular area chosen was the rift valley forming at that time between Australia and Antarctica, as those two continents were in the initial stages of separating in Early Cretaceous times.

Peter also had to deal with the layout Australia Post required. The challenge was to make a credible scene in which the dinosaurs fitted into the rectangles that Australia Post had placed for each stamp in the block. The dinosaurs had to be large enough to fill each rectangle, but perspective needed to be maintained. In order to do that, Peter drew a scene in which the larger dinosaurs were placed in the background and the smaller ones in the foreground.

Yet another painting, showing a living *Leaellynasaura* on a sandbar adjacent to a swollen river in spring, was commissioned by Australia's national airline, Qantas, for

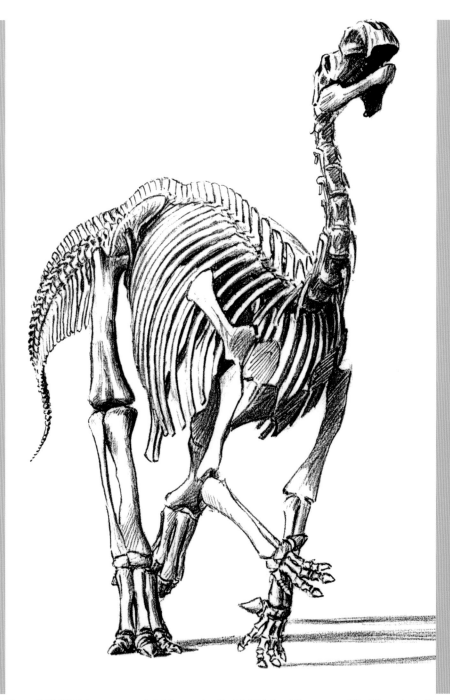

Figure 7.12 *'Muttaburrasurus – skeletal reconstruction'* (1992). Graphite on paper, 28 x 25 cm, Australia Post collection (P. Trusler). Skeletal drawing of *Muttaburrasaurus* reconstructed for the stamp design. The skeleton is more than 7.5 metres in length.

a 1993 article in their in-flight magazine titled 'The dinosaurs that came in from the cold'. At the time, Qantas was paying good writers fees of $3000 per article. Pat and I transferred this directly to Peter for new artwork, and we were delighted to have 12 000 copies of the magazine in which the article appeared delivered to our doorstep! It often remains important to propose such deals because, as in scientific pursuits, the realities of publishing do not always translate into commissions for original artwork.

Later, when the cover for the book *Dinosaurs of Darkness* was under consideration in 1997, the objective was straightforward. This artwork needed to convey, in a single image, the concept of dinosaurs living in a challenging polar environment. This was accomplished by Peter's portrayal of a frigid scene during winter darkness with one dinosaur (*Timimus*) hibernating and another (*Leaellynasaura*) being active in the pale light of the *Aurora australis* (*see* figure 7.1). Fortunately, I had been at high latitudes in the past, both in Antarctica and northern Alaska. Because of these experiences, I was aware of two critical things. First, that for a fortnight each month at these latitudes, the moon goes from half full to full to half full during the polar winter and then disappears for the other half of the monthly lunar cycle, the new moon phase. The other insight I had gained was that the aurora could be quite visible during a full moon. So, Peter was able to paint a scene with a credible amount of light, even though this scene was set in the depths of a polar winter, a time truly without sunlight.

Figure 7.13 '*Muttaburrasurus* – skull' (1992): study from the Queensland Museum specimen. Graphite on paper, 22 x 23 cm, collection of the artist (P. Trusler). Preliminary sketch of the skull of *Muttaburrasaurus* for the stamp design.

By the time the painting was commissioned, two lines of evidence suggested that *Leaellynasaura*, and dinosaurs like her, may well have been active during the winter. The first evidence pointing in this direction was the structure of her brain. Her optic lobes were unusually large for her family of dinosaurs. This could be interpreted in two ways. Either it was simply the normal pattern of growth of the brain, with juveniles of a given species having a relatively larger optic lobe compared to the rest of the brain than adults did. Or it could have been because she had an enhanced capability of processing the weak visual signal coming in from her large eyes during the winter. The juvenile's brain is the bodily organ that is closest to its adult size, so the first explanation was thought to be unlikely. Therefore, it seemed she was particularly well adapted to seeing under the low light conditions of the polar winter.

Another line of evidence, even more compelling, suggested that *Leaellynasaura* had some form of endothermy and so would have been able to actively cope with a polar winter. Nearly a decade after the first relevant insight relating to this matter, Anusuya Chinsamy from the South African Museum, Cape Town, was given a small piece of bone of *Leaellynasaura*, together with one from the ornithomimosaur *Timimus*. When Anusuya cut a thin section of their bones, she observed a marked difference between the two dinosaurs. The bone sample of *Timimus* had successive dark bands in them. Anusuya interpreted these structures as lines of arrested growth (LAGs). In living animals, LAGs represent periods in the life of an animal when bone growth slows down and what little bone is deposited at such times is significantly denser than the adjacent bone, laid down when the animal is more metabolically active. These LAGs can be deposited when the animal does not eat for prolonged periods, or has a reduced intake of water, or when it hibernates. Perhaps *Timimus* hibernated. *Leaellynasaura,* on the other hand, lacked LAGs, suggesting that during her lifetime her metabolic rate never decreased markedly. That meant it was likely that she was never under stress from lack of food or water, and that she did not hibernate. This intriguing hypothesis deserved illustration.

So, the scene for the cover of *Dinosaurs of Darkness* was plausibly set in winter, with the *Aurora australis* overhead and the ground illuminated by a full moon. In this scene it was credible to have a group of *Leaellynasaura* foraging actively for food, much as musk oxen do during the Arctic winter today, while an individual *Timimus* hibernated nearby.

(Following pages) **Figure 7.14** 'Australia's dinosaur era' (1993): reconstruction of the Australian Early Cretaceous. Alkyd oil on linen, 43 x 91.5 cm, Australia Post collection (P. Trusler). Australia Post issued these stamps in 1993, the same year that *Jurassic Park* premiered. The setting for these Australian dinosaurs was the rift valley that then existed between Australia and Antarctica. Peter's challenge was to come up with a plausible scene within the formats specified by Australia Post. Dinosaurs from left to right: *Leaellynasaura*, an allosauroid, *Muttaburrasaurus*, *Minmi*, and *Timimus*. In the background are the small herbivores *Atlascopcosaurus*, while pterosaurs (not dinosaurs themselves) soar in the sky. For scale, *Timimus* stood about 2 metres high.

The Artist: Peter

The Mesozoic piece to feature in *Wildlife of Gondwana* (*see* Chapter 6) involved an unusual story that was unfolding from Dinosaur Cove, along the southwest coast of Victoria. This was to present another range of challenges. Tom, Pat and I proceeded as we had on previous reconstructions, so I was being fed all manner of scientific papers and comparative material. I needed little introduction to the research that Pat and Tom were gleaning from their sites along the Victorian coast, but I needed to assimilate volumes of literature on dinosaurs generally. The associated information I would require on fauna other than dinosaurs and on palaeo-ecology – including data derived from palynology, macro-palaeobotany and even sedimentology – were all at my disposal from the army of Pat and Tom's colleagues, who were also involved in studying the data pouring in from various Victorian Cretaceous sites, including Dinosaur Cove itself.

The same sorts of issues regarding incomplete fossil material described in previous chapters also applied here. Understanding the muscle systems that were thought appropriate for dinosaurs remains a work in progress. All manner of comparative methods also needed to be brought to bear on matters such as eye anatomy, skin structure and pigmentation, posture and the possible behaviour of the dinosaurs recovered from the Victorian Cretaceous sites. For much of this information, I needed to consider the extensive knowledge gleaned from overseas collections. As a result of evolutionary and temporal distances, extending to some 100 million years ago from the present, the palaeobotanical reconstructions would become more problematic. Everything was likely to be unfamiliar to me.

An Australian like myself is used to a landscape of flowering, sclerophyllous trees, despite the fact that the local flora includes some remarkable Mesozoic plants as 'living fossils'. Familiarity and personal preferences are significant factors that guide an artist, but when one's skills are to be applied to novel information about unseen environments those usual advantages can be an impediment. In looking at the historical spectrum of palaeo-art, I could see many instances where a regional, modern landscape had inadvertently and inappropriately been superimposed on the fossil information reflecting times long ago. In some cases, of course, the knowledge concerning possible differences in a past landscape was not available at the time the art was produced. Nevertheless, these factors are still remarkably pervasive, even when the evidence is available to the artist. At the very least, I was familiar with some of the beautiful varieties of leaf fossils from the Cretaceous in Victoria. Researchers were well aware of differences in regional botanical associations and how these sites changed over time in the Early Cretaceous. But the issue I faced was much greater than simply the reconstruction of individual plant macro-fossil species. I had to try to reconstruct their ecology, and visually communicate something of their natural associations and their distribution in the ancient environment that would constitute a 'landscape'.

The blending of all these things is complex and subtle. Most of us are scarcely aware of what might distinguish a familiar environment from one that is foreign or just 'not quite what we are used to'. The differences that we subconsciously register can be due to the quality of light, the percentage of humidity or even unique smells, for example, and only some of these factors have visual signatures. These aesthetic issues can be made to impact on reconstruction illustration and need to be considered, especially when the work is intended to 'popularise' science.

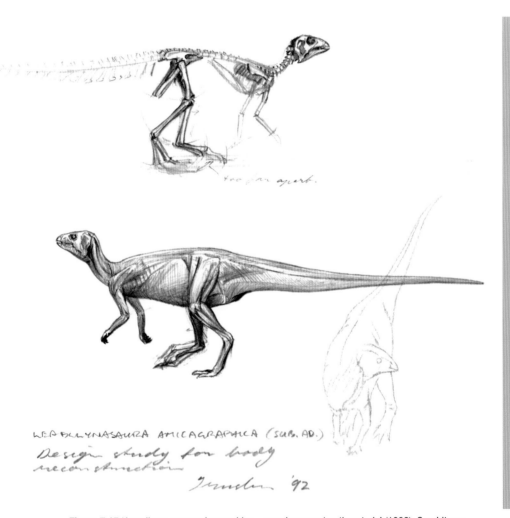

Figure 7.15 '*Leaellynasaura amicagraphica* – muscle reconstruction study' (1992). Graphite on paper, 28 x 38 cm, private collection (P. Trusler). Preliminary sketches of muscle reconstruction for *Leaellynasaura* in the spring river scene. The dinosaur is about 1 metre in length.

To deliberately make information comprehensible and interesting to a wider audience, artists are often encouraged to present their images in a particular way. This generally brings into play both obvious and subtle cultural distortions, or overlays, to the presentation. Some will relate to the time and place in which the artist is working and will be a consequence of the purpose of the commission. Art styles in vogue at the time may take precedence or the art may be stylistically tailored to an intended audience. Other influences will be totally extraneous, but all will be quite unrelated to the time and place being depicted.

In the early years of palaeo-reconstruction art this factor infused works with theological themes. Reconstruction artists employed the same visual figurative and pastoral scenic language as used in theological art because it was already familiar to the wider public. In one sense this was logical, since creation was the immediate cultural comparison that was made to the new science of palaeontology, but the similarity of visual imagery meant that the conceptual gulf between the two was not made clear in the scientific art at its outset. I often ask myself whether the illustration

Figure 7.16 *Qantassaurus* was a small plant-eating dinosaur, a relative of *Leaellynasaura*, about a metre in length. It was named after the Australian airline Qantas for all the in kind support rendered in exhibiting the Victorian dinosaur finds all around the world.

process inflicts injury on the science it is intended to illuminate. Over the ensuing time of scientific discoveries, the culture of impact from mainstream art history abounds. Broadly speaking, the 18th century English landscapes of Capability Brown, the 19th century works of German artist Caspar David Friedrich and German-American painter Albert Bierstadt and, in the late 19th and early 20th century, the American southwest art of Frederic Remington, have all unknowingly made strong stylistic and emotional contributions to palaeo-illustration.

As the various disciplines of research were coming together in this project, and I was contemplating both the details and the broader picture, I began to look at the growing phenomenon of 'Dino-art'. Frankly, I saw some brilliance, but also much banality, despite the astonishing plethora of spectacular species being discovered, and the careful reconstruction and portrayal of many of these. At the time I approached this project, simultaneous shifts in palaeontological interpretation were producing waves of equally stunning revelations throughout biological science. This is not to say that we illustrators had made little, if any, progress. The styles, media and reproductive technologies being employed had been changing quickly. The volume and diversity of the trade was exploding. The inventiveness of some work was demonstrated by a great awareness of contemporary art and commercial advertising. Was it, then, too much to expect some aesthetic sophistication in all this change? I saw some, but not a great deal.

Perhaps my negativity here relates to the approach where an emotional state is imposed on a construct of the past. I am familiar with the concepts of an idealised 'Idyll' or 'Eden' in landscape and figurative art, and can see how these have been developed over the centuries. There is often a powerful belief system incorporated into scenes of the past, where the visual medium invites the viewer to accept that this is precisely the way things were. Invariably, the illustrator makes an emotional appeal to this effect, and it comes in any number of stylistic and cultural guises: the romantic solitude of the wilderness; concepts of the perfection of the natural order; the idea of the 'noble savage' transformed into the 'noble beast'. Intense realism, or finely resolved detail, does not of itself convey truth, but it is so easy to accept that it does. That's just the way we are.

The excitement and competitiveness of the researchers who are making these scientific discoveries and publicising their outcomes have fuelled other emotions too. The illustrations that interpret these discoveries are frequently being produced to grab attention and excite interest. All manner of basic tactics are at the artists' disposal for this, and some of the scientific data has been immensely helpful. Dinosaurs have not only diversified, but have also become more versatile, more capable and much more intellectually calculating! There has been a powerful resurgence of the mostly masculine preoccupation with the formidable and aggressive predatory foe. Dino-art has produced a plethora of fantasy, the likes of which the world has never seen. That, by itself, is acceptable, but I think that a scientist needs to be prudent. For all of the enigmas that the scientific method attempts to elucidate, much of the palaeo-art does not progress beyond who eats whom or which is the victor and which the vanquished.

(Previous page) **Figure 7.17** '*Leaellynasaura amicagraphica* – corpse' (1990): reconstruction of the Otway Basin, Early Cretaceous of Victoria, Australia. Alkyd oil on linen over hardboard, 51 x 71 cm, private collection (P. Trusler). *L. amicagraphica* corpse painting produced for *Wildlife of Gondwana*. The total length of the specimen is about 1 metre.

It says more about the aggressive tribal psyche of the artist and the audience than it does about the intellectual deliberations of biological science – even though there might be a connection.

The 19th century preoccupation with menageries for the cultural and intellectual elite and the concurrent explosion of natural discovery, along with a curiosity about the exotic, have left a continuing legacy in natural history illustration. Understandably, it is alive and well with regard to dinosaurs. This form of exotica has now expanded into outer space, and cultural concepts of the 'alien being' are not only being imaginatively fuelled by palaeontology, but are also returning to illustrate it.

My tentative steps thus far were not immune to my own critique and so I wondered how I should proceed with a dinosaur subject. I killed it! How courageous was that? I shall explain.

The material from Dinosaur Cove, with which I was mostly concerned, had been recovered by Tom and Pat's team from a mined seam that lay buried in the base of a coastal cliff. This was the final resting place of the chosen dinosaur subject, *Leaellynasaura amicagraphica*. This small, herbivorous species was chosen simply because it was the most complete dinosaur known from the fauna of that time – even if represented only by articulated partial skeletons of two immature individuals. There was a small range of other isolated bones known from several different locales, a variety of elements representing a mixture of age classes and closely related species. The sediments surrounding the two partial dinosaur skeletons, representing a quiet, backwater setting, contained little of the other flotsam and jetsam of the time, but this material was known from other places in sediments of about the same age, and from slightly different settings. The four-toed foot anatomy of the little *Leaellynsaura* presented some problems for interpreting the precise stance of the animal, and forelimb elements were lacking. Again I was cognisant of two things. The superb, detailed preservation of these fossils that I could examine from a host of local Early Cretaceous sites was compelling. But being able to put it all together representing one place and one time, and to connect much of it in a full scenic restoration, was both confounding and confronting. Of course, the detail of what was associated with what and when would greatly improve over time with continued collecting, mapping and analysis, but I had to work with what was at hand and do the best I could within the limitations of the data.

It occurred to me that, in fact, putting the bits of this incomplete puzzle together was the central task of my commission. The diversity and pattern of pieces attained greater significance. I could show all that was understood about the immediate environment of *Leaellynasaura*'s last resting place – a moment prior to the burial of its remains and of the material that might have lain about the corpse in nearby subenvironments. I could thereby paint a scene from the past, and at the same time show something of the substance of what palaeontologists really have to deal with in order to unravel that past and decipher the natural processes in operation at the time. The sedimentology revealed how patterns in the clays and sands translated into comprehending specific past environments, and I could show this in my reconstruction too. The distribution of decaying leaf litter could be illustrated, with small carapace fragments of freshwater turtle shell, lungfish teeth and herbivorous dinosaur teeth scattered among it. These plants and animals were not themselves found in the final resting place of the two *Leaellynasaura* partial skeletons, but they were certainly preserved this way in different environments across this broad river system through cyclic deposition.

I was all too aware that the majority of evidence used in understanding the detail of what has happened in the past is not to be found in prime specimens in museum cabinets. Carefully collected field data is a must. In the case of my reconstruction of *Leaellynasaura*, I needed data such as the chemistry and physical nature of the sediments surrounding this material. Microscopic evidence of plant spores and pollen, as well as the taxonomy and morphological nature of the leaves and bark of the fossil plants, impacted on my work. Even the pore structures on plant cuticles and the shapes of the leaves themselves provided insights into the annual stresses these ancient plants had to endure in life. There was dating to be done, sandstone grain structures to be examined and sedimentary layers to be sequenced. Even the analysis of the palaeomagnetic nature of the concretions so common in the rocks associated with *Leaellynasaura* was critical to placing the continental block on which this little dinosaur had lived in its correct geographic position relative to the South Pole of the day. The working reality was the careful coordination of a multiplicity of evidence. My job, working with Tom and Pat, was to sift through all this debris – debris that had undergone momentous changes and survived, against all the odds. This told us so much: because it reaffirmed to us that the prime specimens were only rare trophies of the game, the world of *Leaellynasaura* had a great deal to offer about the process and substance of enquiry.

Hopefully, I could invite the viewers of my finished art to investigate a typical setting at the outset of the fossilisation process. The information about the cool environment and ecology could be represented at the same time. I hoped to produce a painting for contemplation, rather than a picture expressing pathos.

Over the ensuing years, I produced a number of works on the inhabitants from the nearby localities in this Cretaceous sequence for a variety of different purposes. In some cases these reconstructions were intended to present a sense of how the scientific interpretations were advancing, increasing the knowledge about already known fossils or presenting new species. In other cases, this art was to communicate totally new interpretations or innovative conceptual ideas. All too often, scientific data can be interpreted in several ways, some of which are actually contradictory. One must sometimes entertain multiple working hypotheses when not enough data is at hand to sort one from another. Sometimes matters are best left in limbo until more definitive information comes to light – which it does, in some cases, but in other cases the 'right' answer may never be known. This does not really bother a scientist. Expert opinion, too, can vary and the process of debate drives science to new understandings. This Dinosaur Cove site certainly challenged prevailing hypotheses about the biological limits of dinosaurs. Issues about warm-bloodedness and thermo-regulation, nocturnal activity, cooler Mesozoic environments at polar latitudes, were all apparent at the outset of my commissions. Certainly, there were unusual taxa represented, which extended the known temporal and distribution ranges for certain groups and inspired theories about this part of the world being a 'refuge' environment for species that had long been extinct elsewhere in the world.

These factors were often difficult to communicate, because various inhabitants or other elements in the unusual environment that I was painting could well be visually contradictory to a viewer used to the modern world. It was interesting for me to be involved in producing images to reflect a factual reality, in one instance, and others that were to deliberately provoke the viewer to ask: 'Could this really be so?'

The concept I developed for the first *Leaellynasaura* image was destined to be redeployed for another reconstruction I found myself working on from this local Early Cretaceous rock sequence. One of the largest animals whose isolated bones, teeth and scutes continued to be turned up as fossils from the San Remo area to the southeast of Melbourne was *Koolasuchus cleelandi*. This animal was not a dinosaur but the last representative of an amazingly diverse and very successful lineage of amphibians known as temnospondyls. These were mostly bizarre-looking creatures, somewhat resembling the offspring of a hypothetical union between crocodiles and newts! Characteristically, temnospondyls possessed conical teeth which had a convoluted, folded internal structure, known as the labyrinthodont condition. They had well-armoured heads and pectoral (shoulder) girdles, all displaying a distinctive, reticulate ornament. Some were heavily scaled, and others apparently were not.

The group had flourished for more than 200 million years, part of the vertebrate evolutionary progression from an aquatic lifestyle to a terrestrial one. Because of their transitional nature, the earliest members of the group had always been of intense scientific interest. Fins were evolving into limbs. Swimming was converting to crawling. Oxygen needed to be extracted from air, not from water. Vertebral columns reflected this transition, and vertebrae underwent a wide range of morphological experiments in order to cope with the stresses that terrestriality and new modes of locomotion were imposing. It appeared, however, that many of their kind diversified into the new semi-aquatic lifestyle and exploited it successfully for long periods. Whereas some were far more terrestrially adapted than others, some actually returned to the water, more or less reinventing the aquatic lifestyle. Like so many of the temnospondyls, *Koolasuchus* was likely an ambush predator, lying in wait for its prey rather than actively hunting it down.

I think it is fair to say that most of these unattractive amphibians from the past have not received the degree of popular attention that the subsequent reptile radiation has enjoyed. I had been thinking about this for quite some time, and when I was given the opportunity to embark on an ancient amphibian reconstruction in 2002 I knew my aesthetic sensitivity was going to be tested. I had been feverishly working on large oil paintings and wished to continue my interest with work of that scale. I knew that the predicted length of *Koolasuchus* was about 3 metres, and I was delighted that my developing skills could be applied to producing an approximately life-sized reconstruction. I already had a grip on much of the habitat information from previous work I had done for Tom and Pat. All I needed to do was to immerse myself, and *Koolasuchus*, in its appropriate stream.

Again, I chose an autumnal setting. This was partly for aesthetic reasons and partly to communicate the seasonal conditions evidenced for the site. The *Koolasuchus* fossils that Mike Cleeland had discovered were invariably preserved in coarse, gravelly sediments. These were hard rocks that presented significant difficulties in extracting the fossils, even for experienced preparators like Lesley Kool. These same stubborn sediments, though challenging for Lesley, held the clues I needed to attempt a reconstruction. They were most certainly deposited in high-energy stream or river environments that were able to move the sizeable pebbles now bound into the conglomerate rocks. At times there were reduced water flows, which deposited finer sediments, more like those surrounding the *Koolasuchus* specimens.

Knowing all this, I did not want to create an underwater setting, because this environment would be least familiar for the viewer. Such a setting would also restrict my reconstruction of the wider scene and associated biota. Again, I was looking instead for a device that would provide a natural, contemplative approach. The large *Koolasuchus* was probably going to be unusual enough to interest the viewer, but so big that I was wondering how to impart believability to something that was going to be quite foreign to both my eyes and those of the art audience. I needed a device that could also clearly provide a sense of visual scale in this aquatic scene. If I was to depict the scene as a view from above the water surface, how was I going to show what lurked below? It came to mind that if I reconstructed a scene during a period of quieter flow, this might allow me to depict more of the total information from both above and below the waterline. This turned out to be a simple enough solution, but one which, unknowingly, would almost be my undoing. The optical effects of reflection and refraction would surely test my abilities of visual illusion on a grand scale.

Pat and Tom approached their colleague, Dr Anne Warren, who was an international authority on the fossil amphibians. I needed another fast-track education to get me up to speed with all and everything about temnospondyls. Throughout the lengthy stages of production of this work, Anne and her colleagues fed me vast libraries of literature and made available comparative specimens from around the world. She recapitulated the theoretical arguments for me, visiting my studio intermittently to review the form models I had sculpted and the large drawings I worked up. Kat Pawley was Anne's PhD student at the time, and Kat's willing participation and research on biomechanics informed my understanding of functional anatomy. She, too, was sharp, and together Kat and Anne challenged my assessments and championed some of my observations.

(Previous page) **Figure 7.18** *'Koolasuchus cleelandi'* (2004): reconstruction of southeastern Australian Early Cretaceous. Oil on linen, 140 x 240 cm, Queen Victoria Museum & Art Gallery collection, Launceston, Australia (P. Trusler). *The last, last labyrinthodont?* was the title given to the scientific paper describing this temnospondyl amphibian found in the Early Cretaceous rocks of Victoria. Previously another, older Australian temnospondyl had been described as 'the last labyrinthodont?' because it was then thought to be the youngest one that ever lived. *Koolasuchus* may have reached up to 4–5 metres as a fully developed adult.

There were important reasons for this intense input from Kat and Anne, because modern salamanders were the only comparable living analogues we had. These rather elegant, diminutive forms were such distant and highly derived relatives that their value in this reconstruction was likely to be dubious. Pat was determined that I should learn more about these in any case, particularly for a better understanding of the ecological and functional context for my subject. An opportunity came via another project. Pat had taken Tom and me to Russia in order to assist her with her interests in the Precambrian research that was being conducted by the Paleontological Institute (PIN) in Moscow. Tom and I returned from a field season on the Russian White Sea coast, and en route back to Australia we took up an invitation to visit the Asa Zoo in Hiroshima, Japan, where Pat, Tom and their daughter Leaellyn had observed the captive breeding program for the rare Japanese and Chinese Giant Salamanders in 1992. Here I was able to study, up close and personal, the largest living species of salamanders (up to about 1.5 metres long – approximately half the length, and one-eighth the bulk, of *Koolasuchus*). I could sketch individuals at all stages of development and also, delightedly, observe these unusual animals in the wild. This group is totally unknown in Australia – past or present. This visit, therefore, was most instructive. I gained insights into the ontogenetic transformation from aquatic, gill-breathing larvae to skin- and lung-breathing adults. Did it reflect long evolutionary transformation giving rise to temnospondyls in any way? Maybe not! I was able to observe the feeding strategies in a large-headed animal – one with a skull that turned out to be an excellent analogue of the flattened, shovel-shaped skull of *Koolasuchus*. I took stock of the locomotion behaviours exhibited by these highly aquatic and large salamanders in fast-flowing streams, so like the environments inhabited by my ancient amphibian. Back at La Trobe University, Kat dissected the related Chinese Giant Salamander with me in attendance, and I gained first-hand insights into the myology of these living amphibians. How much of the detail of these muscle systems would relate to those of *Koolasuchus*?

When I needed further input regarding the animals that had been the contemporaries of *Koolasuchus*, I was introduced to drawers and drawers in the Museum Victoria – collections of fossil insects and fish, collected years before by a series of expeditions from Monash University and the Museum. The productive fossil locale was Koonwarra, in south Gippsland, which represented a series of ancient, quiet-water lake deposits. Not only were there fish and invertebrates beautifully preserved in these cold-water lake muds, but also plants, some being remains of the very plants that I had detailed in my previous dinosaur images.

Figure 7.19 'Japanese Giant Salamander' (2003): notes and sketches from Asa Zoo, Hiroshima, Japan. Graphite on paper, 22 x 23 cm, collection of the artist (P. Trusler). Some adult giant salamanders can reach nearly 1.5 metres in length.

Figure 7.20 'Japanese Giant Salamander – larval studies' (2003): notes and sketches from Asa Zoo, Hiroshima, Japan. Graphite on paper, 22 x 23 cm, collection of the artist (P. Trusler).

I attempted to select species and their life-stages suited to the habitat and season, placing them in the painting in spatial and behavioural contexts that would likely be natural. The enormous size disparity between the different elements of the biota was not an issue. The reconstruction was to be rendered as a 2.6 metre painting, the centrepiece for an exhibition at the Monash University Science Centre – the premier exhibition to celebrate the launch of its new building. This illustration was not commissioned primarily for publication. A large proportion of the subjects in this large painting were designed to not be immediately seen by the viewer. They were subtle, and often quite small. They were there to be discovered only by close inspection. Some of these smaller subjects were turtles. Turtles were commonly preserved, but the exact taxonomic identity of their fossils, known from both the Otway and Strzelecki sites, was inconclusive at the time of my painting. Meaningful comparisons to possible related forms were nigh impossible, in part due to the incompleteness of the fossils. However, they were such an abundant component of the Victorian Cretaceous biota that I felt compelled to include them, if only by their scavenged and bleached remains.

Figure 7.21 'Summer solstice ephemera 1' (2007): study of the Australian Emperor Dragonfly, *Hemianax papuensis*. Pastel, graphite and gouache on paper, 18 x 32 cm, private collection (P. Trusler). Summer solstice ephemera study concept that was employed for the dragonfly reconstruction in the *Koolasuchus* image (figure 7.18).

Figure 7.22 '*Coccolepis woodwardi*' (2003). Graphite on paper, 28 x 38 cm, collection of the artist (P. Trusler). Cretaceous palaeoniscoid fish posture drawings. The image is close to life size.

Figure 7.23 '*Ceratodus* – reconstruction study' (2003). Graphite on paper, 28 x 38 cm, collection of the artist (P. Trusler). Reconstruction of a juvenile lungfish from the Early Cretaceous. Although quite variable in size, some lungfish may have reached up to a metre in length.

In addition to the turtles and insects, there was quite a variety of freshwater fish, including most major groups that are still alive today. Hundreds of the complete skeletons of these fish had been collected from Koonwarra, as well as a few here and there at other sites along the Victorian coast. From this collection, I chose to represent a young lungfish, *Ceratodus*, and one of the small palaeoniscoid fish, *Coccolepis*. This palaeoniscid species was interesting because it, like *Koolasuchus*, was one of the few survivors of a once much more diverse group – a real relic of the past, even in the Early Cretaceous.

The Koonwarra *lagerstätte* included beautifully preserved specimens that had already been well studied. For example, some of the insect wings even preserved minute hairs, detailed venation that was critical to their taxonomic placement, and even pigment patterns. There were aquatic insects and insect larvae, as well as small primitive freshwater shrimp, all of which could be part of my reconstruction. The Koonwarra sediments were also world renowned for their preservation of some of the oldest 'bird' feathers. While the feathers have thus far eluded precise identification, their presence was certainly indicative of likely primitive birds or advanced theropod dinosaurs living alongside the waterways. The avian presence was later supported by the discovery of the furcula (wishbone) and hind limb bones of primitive enantiornithine birds.

My next choices related to the flora. One of the most beautiful taxa in this Early Cretaceous polar flora was the ginkgo, *Ginkgoites australis*, with its striking, fan-shaped leaves. It was deciduous, like the surviving maiden-hair tree from China, *Ginkgo biloba*. The presence of extensive leaf mats left behind as fossils was indicative of an annual leaf drop controlled by the season. Presumably, this early species also changed colour prior to shedding leaves for winter. I had conceived these leaves as a pattern motif for my painting, because I wished to create repeating visual rhythms throughout the image. These were to be a metaphor for the physical pattern distributions to be found within a landscape, the repeating structural designs that occur throughout biological anatomy and the patterns of change operating through evolution. This theme was to be evident at a reflected level, a surface level, and at a refracted, transmitted level. The layering was another metaphor to me: I wished to create a multi-level image with its inherently generated distortions, reminiscent of the distortions and biases that palaeontologists need to accommodate in the processing of information from a variety of disciplines and analyses, in order to reach a robust understanding of their subject.

My design imposed some technical considerations on the painting as well. Obviously, I was proposing considerable toil for myself, if I was to depict such a detailed scatter, and more so if I was intent on a hard-edged realism. I knew that, technically, this was not going to work to the best effect if I had to paint the intense colour of the leaves over dark reflections on tannin-stained water. Glowing oranges or yellows would tend to brown and green respectively. The solution was to stencil the leaves, but in reverse. By cutting the leaves out of self-adhesive film and placing that film over a bright yellow ground I could paint an uninterrupted dark background over the canvas and then remove the stencils to reveal glowing, sharply defined foliage across the water surface. Detailed modulations for form and optical effects related to the air–water interface could be added later. I constructed the scenic design with linear perspective lines to calibrate the diminishing size and angle-of-view ratio over the canvas. It was important to keep the visually inferred water surface flat. I enlisted my wife Gael and artist Glenda Wise to help with the stencil cutting, and the painting was underway. This was no small task, for either myself or any of my willing collaborators!

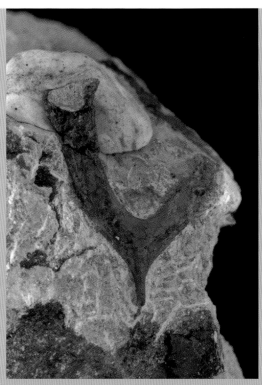

Figure 7.24 Furcula (wishbone) of a bird from near Inverloch, Victoria. Very few bird bones have been found in the polar Cretaceous sediments of Australia, though feathers are known from an inland site near Koonwarra. This little bird, the size of a pigeon, was a member of a primitive group, the enantiornithines. These birds were very successful for much of the Late Cretaceous, only to be crowded out by more advanced forms in the Cenozoic. This wishbone was bitten by either a fish or a small terrestrial vertebrate, as there are four tooth holes in the specimen. The left branch of the clavicle is about 30 mm in length (photo by S. Morton).

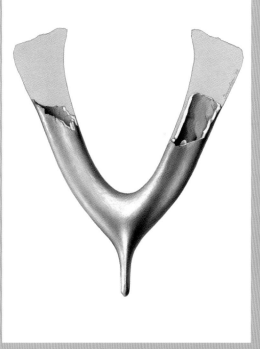

Figure 7.25 'Enantiornithine bird furcula' (2008); standard specimen illustration, P208183, reconstructed dorsal view, Early Cretaceous, Strzelecki Basin, southeastern Australia. Gouache on paper, 24 x 24 cm, private collection (P. Trusler).

8
A Moa Mummy: A classic dissection

Figure 8.1 'Upland moa' (1988): right profile of mummified head, *Megalapteryx didinus*, NZ.B16298, illustrated life size, Museum of New Zealand Te Papa Tongarewa, Wellington. Watercolour and gouache on paper, 29 x 22 cm, private collection (P. Trusler). The Nelson moa's mummified head in its original condition. The distance from crown of skull to base of neck is about 18 cm.

The Scientist: Pat

Many continents today host large ground birds: the ostriches in Africa, the rheas in South America and the emus and cassowaries in Australia. Looking back in the past, there are even more: the giant elephant birds of Madagascar, the menacing, highly predaceous phorusrhacoids of South America (and even North America) and the mihirungs (dromornithids) in Australia.

When I began working on the giant dromornithids from the Cenozoic of Australia (*see* Chapter 9), I kept pondering the relationships of all these huge avians. Clearly the phorusrhacoids were carnivores, and their relationships seemed to be with the gruiform birds, represented today by the cranes and rails – but then, maybe not. The prevailing wisdom for most of the rest of these gargantuans was that they were all somehow related and had a distinctive palate structure – something called a palaeognathous, in contrast to the neognathous condition of most other living birds. This primitive palate seemed to be very immobile, and so had significant functional implications for this group.

The palate of the dromornithids was not to be fully understood until outstandingly well-preserved material representative of this group was discovered at Bullock Creek (south of Katherine, in the Northern Territory, Australia) and other sites, primarily as a result of work done by Peter Murray and Dirk Megirian at the Northern Territory Museum, and to a much lesser extent by Tom and me, working in parallel with Peter at the Bullock Creek site. The cranial material that was acid-etched out of the hard limestones of that site changed the family tree of this now extinct group – instead of being thought of as close cousins to the emus, they were now considered to be close relatives of magpie geese! Science marched on.

But, back to 1985. My thesis finished and published, I had some time to revisit the topic of palaeognathous (ratite) birds, and even to question the reality of this grouping. Clearly the emus and cassowaries were closely related, and there seemed to be marked similarities in bone morphology between the now quite geographically separated rheas of South America and ostriches of Africa. A little less clear was the relationship of the New Zealand kiwis and the now-extinct giant ground birds, the moas. Even this latter group survived until not so long ago; in fact, there were a few mummified moa remains held in two or three museums in New Zealand, and one overseas. I now had access to soft tissue of dinornithids, and it dawned on me that it might be worthwhile to study the moa mummies to extract a great deal more information about at least one group of these giant ground birds than was not possible for any other completely fossil groups. And it might just solve the debate concerning the relationship of moas and kiwis, and other supposed palaeognathous birds.

John Yaldwin, then Director of the National Museum of New Zealand (now Museum of New Zealand Te Papa Tongarewa), was the person who made this study possible. John was willing to let my team and me study in detail his unique 690+/- year-old specimen of one of the smaller moas, *Megalapteryx didinus*. He allowed us to rehydrate and dissect it as well as sample it for DNA and collagen content, check the feathers for the parasitic Mallophaga, or feather mites – in fact, to do whatever was necessary to wrest as much information out of this ancient bird as possible. We were so grateful to be allowed to proceed with this study, for it was impossible to gain access to any of the other specimens – understandably, because of their rarity. As a scientist, it was such a privilege to have been given this access.

The specimen was duly packed carefully, and I carried it in my hand luggage to New York, where my graduate supervisor, Walter Bock, and I slowly rehydrated this long-dead 'chook', which we called Oliver (in honour of Dr W. R. B. Oliver, then Director of the Dominion Museum in Wellington (now also part of the Museum of New Zealand Te Papa Tongarewa), who had discovered our specimen in 1943. Before this, however, we had taken small samples that would be analysed for the DNA and collagen content, and examined for possible Mallophaga. Mallophaga are quite group-specific, and so we hoped that would tell us something about the relationships of our moa. None were to be found, however, which was a disappointment. There was degraded DNA present that appeared to have belonged to our now extinct moa – not bacterial in origin – and it reacted strongly to the living kiwi DNA. In subsequent experiments carried out by Roger Cooper's group, segments of the *Megalapteryx didinus* mitochondrial genes were amplified, and sequence analysis carried out. Those analyses on the mitochondrial 12S ribosomal RNA suggested that this late surviving moa belonged to a monophyletic group of moas and other ratite groups, but that it was surprisingly distant from the kiwis. This suggested a separate and lengthy evolution for the moas and the kiwis. Degraded collagen was also present, and it had retained its helical configuration, exhibiting much of its original immunological reactivity.

For six months, the moa mummy was soaked in a rehydrating brew (50% glycerine, 25% ethanol, 23% distilled water and 2% phenoxyethanol) with a series of changes of these liquids. The specimen was kept refrigerated at all times, other than when dissection was underway. At the end of this process, Walter and I ended up with a glycerin- and ethanol-laced head and neck of *Megalapteryx*, as soft and pliable as that of a turkey just beheaded for a Thanksgiving feast. And so began our careful dissection of this not-so-long-dead mummy, working on only one side of the head so that the other side remained untouched and intact.

Once the dissection was complete, and the sketches and photographs – made at each step of this delicate and meticulous dissection – had been collated, the bird was essentially put back together. None of the nine cranial muscles we were able to isolate had been removed, only cut through in order to gain access to the deeper tissues and examine their muscle architecture. The skin was also put back in place, and the specimen was ready to be returned to Wellington, New Zealand – this time not as a dried out mummy, but preserved as a wet specimen like a modern species.

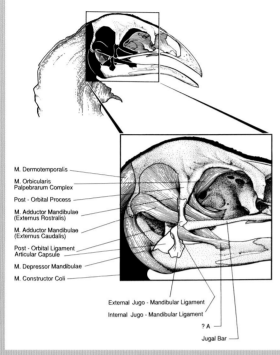
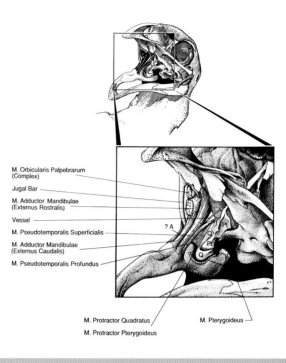

Figure 8.2 Line diagrams of the dissection of the rehydrated *Megalapteryx didinus* from Nelson, New Zealand. Abbreviations: M. – muscle (P. Trusler).

Our dissections clearly illustrated that the *Megalapteryx* musculature and ligaments were quite distinct in their arrangement from those of the kiwi, and as noted, later studies by Roger Cooper and his colleagues further supported this distance in relationships when they analysed the DNA extracted from other moa specimens. Even though our *Megalapteryx* certainly had some DNA fragments preserved as well as collagen, the DNA was not complete enough for determination of relationships. The collagen was sufficient to allow us to date our specimen at around 600–700 years old with plus or minus factors of 95–120 years – so it was certainly a bird that had overlapped in time with the earliest humans to colonise New Zealand.

In the process of this dissection, we had removed a tiny bone, the stapes, from the ear of the mummy. In birds, unlike in humans, there is only a single ear bone. Humans have three, one being the stapes. This little bone transmits sound vibrations from the eardrum to the inner ear, and in birds the shape is characteristic in each taxonomic group. It was photographed and placed in a separate container to be transported back to New Zealand – but all of this material made an extra stop on its way home. I retained it for a time, with the permission of the National Museum of New Zealand, at Monash University, where Peter made a series of illustrations. It was clear from our

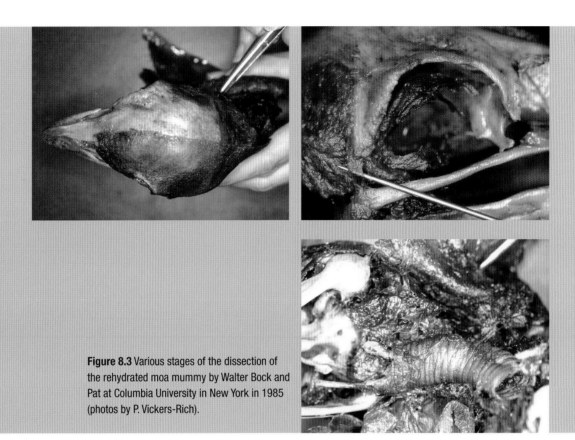

Figure 8.3 Various stages of the dissection of the rehydrated moa mummy by Walter Bock and Pat at Columbia University in New York in 1985 (photos by P. Vickers-Rich).

examination of this tiny bone that it showed little similarity either to the living kiwis or to another primitive bird group, the tinamous, thought to be ancient relatives of the ratites. In fact, it was unusual even in the group called ratites: it had a natural kink in the middle of the shaft, a small basal fossa and no struts, among other characteristics (see figure 8.3). There was some similarity, however, between the stapes of *Megalapteryx* and that of the living Australian Emu and the African Ostrich, indicative of a relationship with some ratites. The shaft of the stapes joined the footplate over a broad area, much more like the structure characteristic of *Dromaius*, the emu. But it was a surprise to us all that there was not more similarity with the co-habiting kiwis – again, supportive of the conclusions drawn from the DNA/RNA analyses. The relationship of the kiwis and the moas seemed to be very distant, and evidently these two groups had gone their separate evolutionary ways for a very long time.

Besides making the detailed sketches of the tiny ear bone of *Megalapteryx* and those of other ratites, Peter concentrated on the stages of our dissections, beginning with a beautiful full-colour rendition of the bird before the research. He made a series of black and white sketches of the muscles that Walter and I had so painstakingly explored during our dissection in New York – layer by layer – based both on what he could see in the specimen and on our photographs and sketches made during that dissection.

Still, I wanted Peter to gain a broader understanding of moas in general and also to observe the plumage of *Megalapteryx* and other taxa housed in several museums in New Zealand, including the National Museum in Wellington, the Canterbury Museum in Christchurch and the Otago Museum in Dunedin. My aim was to have Peter also speak with a number of moa experts in these institutions – people like Trevor Worthy and Ewan Fordyce. This would give him a broad perspective on the ideas that each of these researchers had on moa interrelationships and their relationships to other birds.

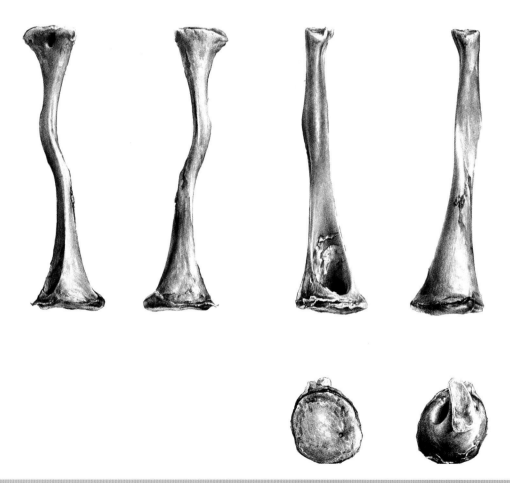

Figure 8.4 'Upland moa stapes' (1988): specimen from Museum of New Zealand Te Papa Tongarewa, Wellington. Graphite on paper, 29 x 22 cm, private collection (P. Trusler). The middle ear bone that links the tympanic membrane to the inner ear from the dissected *Megalapteryx* cranium. Specimen depicted in standard views; left to right: lateral, medial, anterior, posterior, dorsal and ventral. The length of the stapes is 6 mm.

It would also give him a better idea of the settings in which they had lived out their lives. So, the funding was found, and Peter travelled to New Zealand to complete his data gathering. The results of this trip were more illustrations: detailed renderings of the individual feathers of *Megalapteryx* as well as another mummy – this time the hind limb of *Megalapteryx* from a different locality than the one from which the skull was collected. This additional specimen was housed in the Otago Museum in Dunedin. The hind limb had preserved skin, ligaments and feathers. Although at the time Peter and I had planned to produce a full-blown reconstruction of the moa in its environment, other projects intervened, and to date it still remains unfinished – when we both have a bit of spare time and I can find some additional funding, we hope we can finalise it!

Peter's reconstruction art appeared first in our research paper on this project, but later resurfaced in our 1999 revision of *Wildlife of Gondwana* as well as being on show in two exhibitions developed by the Monash Science Centre. The original reconstructions were donated to the Queen Victoria Museum and Art Gallery in Launceston, Tasmania, to which Tom and I have continued to donate much of Peter's art, due to the long and productive friendship and cooperative working relationship with Chris Tassell, who was director there for nearly three decades. The QVMAG has long been outstanding in its appreciation of and high quality preservation of art, including a large collection of what we call 'scientific art'. It therefore seemed the place that would be most appreciative of the content of these meticulously accurate palaeo-reconstructions. And today, under the directorship of Patrick Filmer-Sanky, the tradition continues.

The Artist: Peter

Illustration work for the purpose of visually recording information has been viewed as the prerogative of photography throughout my lifetime. So, for me, drawing for that purpose has largely been my process of discovery and learning, and to jog my memory about the detail of what I am viewing. It has advantages in some respects, and disadvantages in others. Occasionally, I am asked to use my skills to depict a unique specimen instead of reliance being placed on photography, and this was certainly the case with the Nelson moa mummy.

Investigations conducted by Pat, on ratite birds in general, led her to study some remarkable mummified specimens of the extinct upland moa, *Megalapteryx didinus* of New Zealand. The living birds had narrowly escaped the attention of European science, only because they had not survived the years of the first human occupation. The hunting pressures and habitat changes that transpired with the first settlers were considerably more subtle than those following European invasion. While the exact mechanisms for moa extinctions may never be precisely understood, there are amazing records of both moa and kiwi feather clothing.

Megalapteryx was the smallest of the giant moas, which include probably the tallest of all known birds, a species of *Dinornis*. But *Megalapteryx* may also have been the last surviving member. It was adapted to the colder and more remote mountainous spine of New Zealand's South Island. Its fossil bones, like those of the gargantuan species, were relatively common, but its mountain habitat also ensured that some specimens had been mummified. Birds had died in upland caves, and parts of their bodies had literally been freeze-dried by cool, dry air flows before the normal processes of decay could strip their bones clean of flesh.

Pat was to take the unusual step of rehydrating one of these specimens in order to dissect a side of the bird's face for anatomical study. An additional test for intact DNA and collagen was also to be carried out. Unfortunately, contamination from decay prevented ample DNA sequences from being obtained for further study, but collagen was recovered, and studied by Merrill Rowley at Monash University.

Figure 8.5 'Upland moa – feathers' (1988): feathers attributed to *Megalapteryx didinus*, Museum of New Zealand Te Papa Tongarewa, Wellington. Watercolour and gouache on paper, 32 x 23 cm, private collection. The length of the main shaft of double-shafted feather on lower right is about 7.5 cm (P. Trusler).

Before the dissection was carried out, I was to record some of these gruesome rarities. I drew and painted a series of images directly from each specimen, so that I might understand them from any point of view. The main reasons for doing this were that both the realism that might be required and the information thus recorded are often more instructive than the standard profile, etc. I have found that, particularly when it comes to 'animating', or depicting the behaviour of a creature, a full comprehension of the form is imperative. Often, the ultimate depiction requires an unusual or a specific angle of view that may not have been predicted at the outset. I had all manner of details – muscles, tendons, leg skin and scales, traces of feather tracts and papillae, facial vibrissae, individual feathers and entire patches of plumage to study. New Zealander Trevor Worthy had a wealth of knowledge and experience to offer about all of the moas through his collecting and research, and he took me to forest habitats to instruct me on the relevant botany as well. By the time the reconstruction drawings were started I had, therefore, produced a variety of works, the purposes of which spanned the basic drawing genres. I had detailed watercolours to document the surviving pigment colours as accurately as my art would allow. While any matter to do with colour is subjective, its record for my own personal use of the original art is essential for my reference. Using any other method to record colour always needs to take into account the technology used and its reproduction. The result still ends up being subjective.

The other aspect is, of course, the detailed recording of physical form, with all the advantages that I outlined at the outset of the chapter on *Diprotodon* (Chapter 3). In many respects, this process holds great advantages over photography, because it is one that needs to transform this record for the next stage. It is not a universal record but a personal one.

The other works were also ones of primary record. These took the form of anatomical, graphic diagrams, and while not intended to record everything, were taken for the purposes of clear, illustrative 'shorthand'. Such ink works as those for *The Fossil Book* are a standard stock-in-trade for easy, cheap and reproducible illustration under all conditions. Other anatomical studies that I rendered, such as those for the stapes (a middle ear bone), were fully tonal to record both dimension and detail, and therefore required higher quality reproduction. Each of these different illustrative pieces was produced with one or more specific functions and outcomes in mind.

The graphite sketches of the specimens from many aspects are again a form of record but, as the name implies, are not detailed renditions. These represented my rapid recording techniques, meant to give me a sense of the overall – an impression,

Figure 8.6 'Upland moa – leg' (1988): left profile of mummified left leg, *Megalapteryx didinus*, Otago Museum, Dunedin, New Zealand. Watercolour and gouache on paper, 29 x 22 cm, collection of the artist. The leg length is 90 cm (P. Trusler).

or guide, if you will. The importance of this was that, while these lack the precision of photography, which I also used for this purpose, they do impart a greater appreciation for the dimensionality of objects. This is something that is crucial to the communication of the human perception of form – I cannot overstress the issue of representing three dimensions in two. They also carry more of an emotional record of my thoughts and responses to the subjects on which I am working. The end result of quite a detailed painted reconstruction needs to retain or represent the entire complex of my knowledge, the state of the science and also my personal appreciation of the subject.

All was ready for me to begin, and I started the reconstruction drawings, literally from the ground up. The detailed botany and environmental information was to feature, firstly with my reconstruction of the upland moa's feet. To me, these encapsulated the specialised adaptive features that intimately connected these remarkable birds to the New Zealand landscape – massive legs, cloaked by feathers to protect these birds in the cold mountainous terrain that no longer echoes with their calls.

I was then to set about formalising the form of the body and of the head. Later, the full synthesis of work regarding these remarkable birds, incorporating everyone's input, was intended to result in the final painting. So far, other projects have intervened and this work is still waiting in the wings, yet to be accomplished.

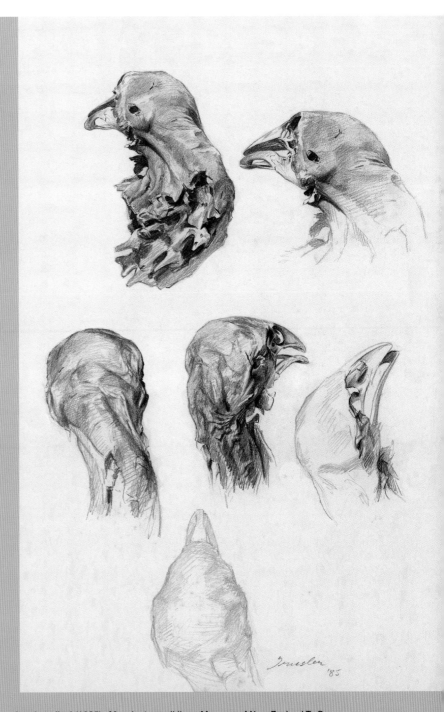

Figure 8.7 'Upland moa – head studies' (1985): *Megalapteryx didinus*, Museum of New Zealand Te Papa Tongarewa, Wellington. Graphite on paper, 38.5 x 28.5 cm, collection of the artist (P. Trusler). Sketches of the original condition of the Nelson specimen, depicted in oblique views.

9 Magnificent Mihirungs

Figure 9.1 Detail – '*Dromornis stirtoni*' (2002): reconstruction of Central Australian Miocene. Oil on linen, 71 x 96.5 cm, Monash Science Centre collection (P. Trusler).

The Scientist: Pat

My first memory of Ruben Arthur Stirton was of a short guy walking down a hall with a tiny mammal peeking out of his white lab coat pocket – a little marsupial that he had brought back to the University of California from Australia. The hall was in the Earth Sciences Building, which housed the Museum of Paleontology, high on the hill of the UC Berkeley campus where both Tom and I were students. Later, in courses with Dr Stirton, and in conversation, he told us students of his work in Australia, which began in the 1950s and continued virtually until his death in 1966. This work built on the achievements of such intrepid explorers as J. W. Gregory earlier in the 20th century, when he used camels instead of 4wd Land Rovers and did not have the backup of the Royal Flying Doctor Service in case of disaster. He and his Melbourne University undergraduate students somehow managed this impeccably, with not a single mishap, in the middle of the austral summer!

Figure 9.2 Ruben Stirton (right) and Paul Lawson on a joint University of California–South Australian Museum expedition into the Lake Eyre Basin in the 1950s. It was Stirton who drew Pat and Tom into research on Australian fossil vertebrates (photo courtesy of R. Stirton).

All of us Berkeley palaeontology students were fascinated by Stirton's work. Some of his graduate students, like Mike Woodburne, had followed up on Stirton's earlier ventures by excavating at sites like Alcoota in the Northern Territory of Australia. There, large numbers of late-Cenozoic, fair-sized marsupials, and even some gigantic birds, had been discovered. Many of the plaster jackets of these as-yet-undescribed Australian fossils were being prepared by a number of enthusiastic staff and students in a large palaeontology laboratory on the ground floor of the Geology Building on the Berkeley campus.

I was completely fascinated by all this activity, not just as a palaeontologist but also as a person with some interest in fossil preparation. I was already working part time, even in my second year as an undergraduate, on curating both fossils and recent material for one of the senior staff, Dr J. Wyatt Durham, who had recently returned from the Galapagos Islands. This gave me training in organising fossil material, and I really thought gaining the further skill of preparation would be of some value. Although I did not have the chance to participate in the actual preparation of the Alcoota material, just watching others do this gave me enough background that later, as a PhD student at Columbia in need of some part-time work, I was able to work as a preparator exposing and preserving some of the higher vertebrates collected on the Tedford expedition, in which Tom and I had participated in 1971 (*see* Chapter 2).

Discovered during the expeditions that Stirton and his students (including Dick Tedford) had carried out over nearly a decade and a half was a reasonably large collection of diprotodontid marsupials and large ground birds whose affinities were not yet clear. The avian fossils were being worked on by Alden Miller, a well-respected and senior ornithologist in the Museum of Vertebrate Zoology, based in the Life Sciences Museum across the Berkeley campus from our building. Alden was the son of the famous Loye Miller, well known for his work on fossil birds of North America and now, still active in his 90s, based at the Davis campus of the University of California. During one semester I had commuted the two-hour drive to Davis to take a functional comparative anatomy course from Milton Hildebrand, the doyen of this field, and after class I would wander into the office of Loye Miller to discuss palaeo-ornithology. He was bright of mind and full of new ideas. I found myself being more and more attracted to this field. What finally clinched my determination to work on fossil birds was a practical class in vertebrate palaeontology given by Joe Gregory on the systematics of the fossil birds from the Pleistocene La Brea tar pits of California. I remember staying back for three or four hours in the lab, by myself, just sorting through the masses of tar-soaked bones to work out just how many vulture types could be identified.

I then took up a fourth year project with Gregory, describing a beautiful and complete skeleton from the Pliocene of southern California – the bones of an unusual scavenging bird, a vulture, with no affinities to the cathartid vultures that now dominate the North American continent. My fossil specimen belonged to a form called *Neophrontops* – a close relative of the Old World vultures, the Gypaetinae. This group had a long history in North America, and their bones were abundant in the fairly recent (geologically speaking) La Brea deposits. Now restricted to Europe, Africa and Asia, the gypaetines are no longer part of the North American avifauna. I realised then that the fossil history of birds could point to a very different picture of the origins of continental avifaunas than that indicated by the living avifauna.

At about this time, most unfortunately for me and for the ornithological community, Alden Miller died unexpectedly. Stirton had discussed with me the prospect of working on the big birds from Australia, with him and Alden as my supervisors. With Alden's death, I was left to work on these, having Stirton as my sole supervisor. Shortly after that Stirton himself died, and this left me alone facing a large collection of fossils, as well as dealing with the loss of two mentors whom I had greatly respected.

When Tom and I decided to seek our PhDs at Columbia, on the other side of the United States, I queried whether I could take the Stirton collections of birds with me to the American Museum of Natural History (AMNH) and Columbia University. This was a big ask, for it was a very large collection, including not only the dromornithids but all of the remaining smaller avian fossils as well. The staff at the Museum of Paleontology gave their permission. Fortunately, Walter Bock, a well-known avian morphologist and functional biologist, was willing to support me in this endeavour. Walter was a careful and meticulous thinker, and happy to spend long hours discussing anything scientific. He became not only a willing mentor, but also a lifelong friend. At the AMNH and Columbia, I had contact with a number of vertebrate palaeontologists working on a wide variety of research programs. And, I was lucky to be assigned a large space at the AMNH where I could spread out my gargantuan bones. Even more luck: a scientific paradigm shift was underway that set the continents in motion, and this dynamic eventually led me to challenge the *status quo* concerning origins of the Australasian avifauna (*see* Chapter 2).

In fact, my mentor Bock had been a student of an icon of evolutionary biology, Ernst Mayr, then based at Harvard, who together with G.W.H. Stein wrote the seminal work on the origin of the Australasian avifauna, *The Birds of Timor and Sumba*. Mayr and Stein's work was set against a backdrop of a stabilist Earth, when continents were not thought to have moved, despite the work of Alfred Wegener and Alexander du Toit. Mayr and Stein's research monograph was published on 11 July 1944. This turned out to be extremely ironic, for that was the very day, month and year I was born – the young upstart who would challenge their ideas, simply because the underlying principles of geology had evolved. It was not because their reasoning was flawed – it was just that the world seemed to have changed!

As I progressed with work on my PhD thesis, the ideas of continental drift and later of plate tectonics came into near full acceptance. This resulted in the recognition that the continents – which Mayr and Stein had thought so permanently placed – had, over time, moved, with consequences for traditional biogeographic interpretations. Quite suddenly, Australia was no longer thought of as an isolated continent, to be settled only by the waves of birds moving into a biological void over 150 million years, but as a continent with a somewhat 'schizophrenic' past. Connections with the great southern continent of Gondwana during the early stages of avian evolution left their imprints in the modern Australian avifauna. Later, when this austral continent broke away from Gondwana, it remained isolated for 40 or more million years, moving slowly northwards.

Only in the last few million years has Australia approached Asia. This impacted on the place of origin and later development of the bird fauna of Australia. Those new realisations formed one of the two themes of my thesis. I was most impressed when both Mayr and Bock took this new view of the world and my new interpretations on board, as an expected outcome of what they and I considered standard practice for a robust scientific investigation. Both accepted my new hypothesis.

A second and major part of my thesis dealt with the study of one group of Australian fossil birds in detail – the large ground birds that Stirton, his students and colleagues had amassed over the years. I began this study, after having passed my qualifying exams in the last year at Columbia, by spreading the entire lot out in the attic of the American Museum, right next to the tower room inhabited by Morris and Marie Skinner, two seasoned collectors for the Childs Frick Laboratory housed at the AMNH. These two characters became my 'scientific parents', mentoring both Tom and me the entire time we were in residence at the AMNH. Along with my supervisor, other senior staff – Malcolm McKenna, Dick Tedford and Bobb Schaeffer (whom we called '3 Bs') – were our intellectual mentors, not just on bird anatomy and biogeography but also on the big ideas and meticulous techniques needed to carry out professional palaeontology. Chester Tarka, the pedantic AMNH photographer, taught me all I needed to know, and more, about how to best image the variety of bones I was faced with in my thesis work.

Figure 9.3 Quinkin figure in a rock shelter on Cape York Peninsula, northern Australia. These gigantic creatures featured in Australian Aboriginal mythology. Such large bird figures alongside the more emu-sized art suggest that the last of the dromornithid birds, the mihirungs, may well have been seen by the early human occupants of Australia (photo by P. Trezise).

So began my comparative work on a group that I subsequently studied for nearly four decades. And this was both exciting and frustrating. I described, bone by bone, a variety of postcranial elements in the collection, which itself was large. In doing so, and taking into consideration the dromornithid material from a number of sites – ranging from the avian scraps then available from Riversleigh, in north Queensland, to the hugely abundant samples from Alcoota – it became clear that there was not just one but several species of dromornithid. Some were deep-billed giants like one that I named *Dromornis stirtoni*. Others were much more gracile, perhaps even cursorial, forms like '*Ilbandornis*' *lawsoni* and possibly even a form known since the 1890s, the goose-like *Genyornis newtoni*. I was delighted to be able to give to each of the new forms names honouring the people who in some way had contributed to my being able to work on this material: those who had organised the digs, or funded them, those who had collected the specimens. I also used the new names I proposed to reflect some aspect unique to the new form. For example, *Bullockornis planei* was a bird (*ornis*) as big as a bull (*Bullock*). The species name was in recognition of Mike Plane who, no small man either, was based at the Bureau of Mineral Resources and had been instrumental in discovering and working this site in the Northern Territory. My teacher and mentor, who had given these very bones to me for study, was honoured in *Dromornis stirtoni* – possibly the heaviest, bulkiest (largest, really) bird that had ever lived, the grand-daddy of them all! The diminutive Stirton, I am certain, would have been amused with this name, had he still been alive.

Figure 9.4 Tom excavating at Bullock Creek in the Northern Territory, Australia. This was no easy task and was carried out mainly with drills and hand tools (photo by P. Vickers-Rich).

Over the months that followed, I sorted out all the different morphologies that I could, keeping in mind the sort of variability one observes in modern species. Because I had very little cranial material, I had to rely almost completely on the limb bones, the shoulder and pelvic girdles and, to some extent, the vertebrae of this mixture of taxa in my attempt to determine the species diversity within this group – one totally unique to Australia. And the other challenge I had to face was trying to work out to what other group these giant ground birds were related.

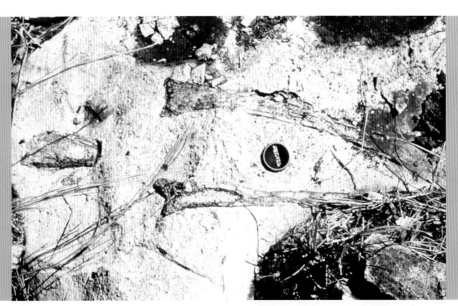

Figure 9.5 Bones of dromornithids, more than 20 million years old, in the limestones of Bullock Creek in the Northern Territory, Australia. 35 mm lens cap included for scale (photo by P. Vickers-Rich).

I sorted and resorted the bones and found that the most frequently well-preserved bone – other than, perhaps, the numerous vertebrae – was the femur, or thigh bone. Vertebrae, however, looked much the same from one bird to another, and were really only distinguishable into size categories – the big ones from Alcoota apparently belonging to *Dromornis stirtoni*, the giant of this assemblage. So, I zeroed in on the detailed morphology of the femora of these birds and used the differences, still taking account of expected variation within a species, to define my different taxa. I found many of the same categories reflected in some of the lower limb bones, especially the tarsometatarsus, but in my sample femora were simply the most plentiful, best preserved and most distinctive.

What I truly lacked and so wanted to find were skull remains, for these would have helped me sort out the relationships of these dromornithids – as, in fact, they did much later when better material was discovered. When I began my studies in the mid-1960s, only one damaged skull, anywhere near complete, existed: that of *Genyornis newtoni*, collected by the Stirling and Zietz expedition from the South Australian Museum in the late 1800s. That skull certainly was the worse for wear when I first examined it, having been looked over, moved, and encased in a plaster 'brick' over decades. When I first saw it, I could do little with it. Missing, critically, was the palate – an area important in working out avian relationships. This specimen clearly brought home to me how important it was to carefully record the detail of a specimen immediately on its discovery. Photography has its place, and this skull had been photographed, but detailed art like that rendered by Peter would have served a good purpose, had it been produced in 1896! Despite the wear and tear imposed on this specimen, the art would have survived and been useful in later phylogenetic studies.

At the time I finished my thesis, my thinking was that the dromornithids were likely related to the cassowaries and emus. So, the reconstruction art provided by artist Frank Knight for the book *Kadimakara*, which my colleague Gerry van Tets and I wrote in 1984, depicted *Genyornis newtoni* in the colour and stance of the living emu. However, over the years, as two of my colleagues, Peter Murray and Dirk Megirian,

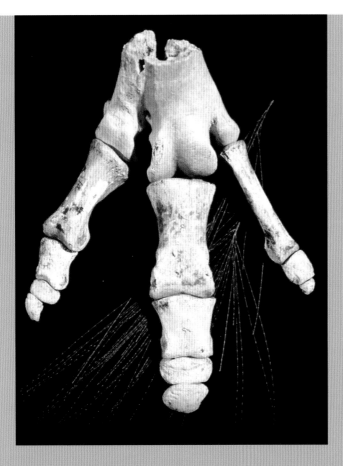

Figure 9.6 The foot of *Genyornis*, the last member of this once successful group of large ground birds on the Australian continent, where large birds were one of the major herbivores. The length of the middle toe is about 15 cm (photo by S. Morton).

excavated further at Alcoota and Bullock Creek in the Northern Territory and Tom and I carried out related excavations, surprising new relationships emerged for the dromornithids. Peter and Dirk on the one hand, and I on the other, came independently to the same conclusion: the dromornithids were not emu relatives at all; their closest alliances were with the anseriforms – a group including the ducks, geese and swans and the South American screamers (the Anhimidae). In fact, once several partial skulls of dromornithids were recovered from Bullock Creek and other more fragmentary specimens from Alcoota, showing the palate and detailed structure of the basicranium, the question of relationships was settled. The dromornithids most closely resembled those structures in primitive anseriforms such as the Magpie Goose *Anseranas*, well known in the swamplands of Australia's northern monsoon country.

And so, with the decades of exploration, discovery and application of a novel preparation technique (acid etching), much greater detail of these enigmatic birds was at hand. Particularly critical was material we collected from Bullock Creek in the Northern Territory that had been encased in solid limestones for more than 20 million years – safely preserving incredibly delicate structures.

Peter was presented with this rich array of materials from which to begin his palaeo-reconstruction work. And with the keen observations of my colleague, anatomist-*cum*-palaeontologist Peter Murray, working with an expansive new collection he and his mates had made – remarkably more complete and better preserved than anything we had had previously – Peter embarked on his reconstructions. First there were pencil sketches. From there he produced the large *Dromornis stirtoni* oil painting (*see* figure 9.12), which graces the front of the book Peter Murray and I (mainly Peter) put together, *Magnificent Mihirungs*. This reconstruction also travels, as an original centrepiece, with the *Wildlife of Gondwana* exhibition. The vibrant red, black and white colouring of the *Dromornis* plumage clearly reflects our collective scientific hypothesis that this avian group is derived from primitive anseriforms – the colours tying to the living magpie goose. So, we colour-coded the family tree (phylogeny) of this enigmatic group of birds! Peter's chosen composition of the birds and the food they were eating (fruits of the vine forests still represented in isolated patches of the Northern Territory) clearly reflects the meticulous incorporation into this reconstruction of all the palaeontological and geological data gathered from those sites yielding the major collections of dromornithid fossils.

The path to Peter's magnificent reconstruction of the *Dromornis stirtoni* has been long and convoluted and will continue into the future. If new discoveries demand, Peter will reflect that in altering his oil on canvas. It will not be the first time he has revised!

(Following pages) **Figure 9.7** Revised reconstruction of *Megalania* and *Genyornis* showing the changes to the head of *Genyornis* that were incorporated digitally to a scan of the original artwork (shown in figure 6.8). Watercolour and gouache on paper, 34.5 x 53 cm. Digitally modified to reflect changed ideas about the relationships of these birds, 2007.

The Artist: Peter

The close of the millennium saw me again focusing my attention on birds. Peter Murray and Dirk Megirian had been meticulously amassing vast quantities of fossil material for research into the origins of, and evolutionary path that had been followed by, Australia's enigmatic mihirungs. Peter had a suite of well-considered conclusions to offer as a result of their toil. Pat's long involvement and experience with this group, too, had led to Peter and Pat's collaboration on a book about the subject. Peter's skills as an anatomical illustrator were formidable, and so it was an enormous honour to be invited to contribute the cover art for their publication. In fact, when I found myself in Peter's laboratory in Alice Springs, I felt somewhat redundant. In the midst of an enormous array of fossil specimens there were Peter's diagrams, working sketches, anatomical models and drawer after drawer of unsorted piles of stunning anatomical studies and reconstructions of a myriad of marsupials and birds. Peter was rather blasé about it all, but to me this was a goldmine, and I suspect that he and Pat knew it too. Peter enthusiastically guided me through it all. Pat and he kept discussing innumerable details about the specimens and the book as he went from bone to diagram and back to bone again. Clearly, they both felt that I could contribute further to the science and presentation of it, and so I got to work, sketching key specimens.

Another of Peter Murray's colleagues was the botanist Peter Latz. He, too, was to become fundamental to my synthesis of information for the final art. I was heartened by the seriousness with which everyone took my possible contribution as Peter M., Pat and Peter L. carted me on a 1500-kilometre (800-mile) ecological tour from Alice Springs to Darwin. Yes, another education was commencing for me as I was walked through the floristics and structure of the ecocline from arid to tropical habitats. The subtlety of the small patches of remnant climax and fire-refuge vegetation en-route was essential to my grasping the trends that may have occurred in the landscape over the past 15 million years. These landscape changes were pivotal components driving the evolution of the fauna, and my companions were keen for me to experience this and then express it visually.

How could I best do this? I anticipated encountering visual issues with such elephantine flightless birds, and Peter M. and Pat had strong preferences for the largest mihirung species from the Late Miocene, *Dromornis stirtoni*, for the featured reconstruction. This choice calibrated any botanical or associated faunal information to around 8 million years ago.

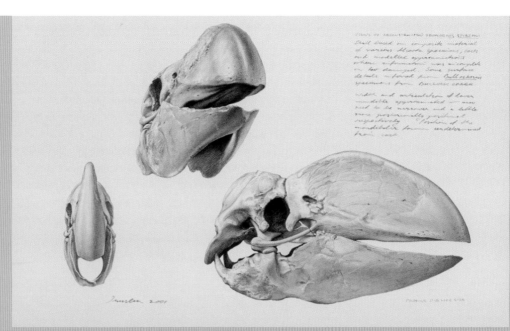

Figure 9.8 '*Dromornis stirtoni* – skull reconstruction studies' (2001), Museum of Central Australia, Alice Springs, Northern Territory. Graphite on paper, 38 x 52 cm, Queen Victoria Museum & Art Gallery collection, Launceston, Australia (P. Trusler).

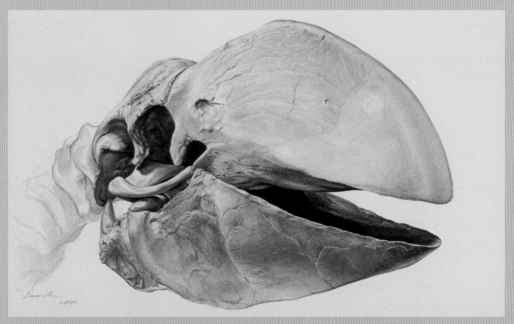

Figure 9.9 '*Dromornis stirtoni* – skull, oblique view' (2001), Museum of Central Australia, Alice Springs, Northern Territory. Graphite on paper, 38 x 52 cm, Queen Victoria Museum & Art Gallery collection, Launceston, Australia. The skulls of D. stirtoni reached upwards of 50 cm in length (P. Trusler).

But *Dromornis* was to be cover art, essentially advertising art for a book that was one of involvement – with enormous detail about functional anatomy, complex theoretical argument and hypotheses. How could I best encapsulate all this faithfully and still fulfil the cover's role?

The focus of the investigation about the origins of the mihirungs had centred on their skull anatomy. This had been the stumbling block to the work on *Genyornis*, but the new material from older species was finally beginning to reveal some answers. The Bullock Creek site, in particular, was providing beautifully preserved and fully three-dimensional specimens that could be acid-etched almost in their entirety from the enclosing limestone. The Alcoota site was providing large samples of younger material, also reasonably preserved. The details of cranial features could now be documented in a number of species, and their similarities to primitive waterfowl could be seen. I now had meaningful modern analogues, detailed osteology and substantial biomechanical interpretations from which to reconstruct anatomy, integument and behaviour.

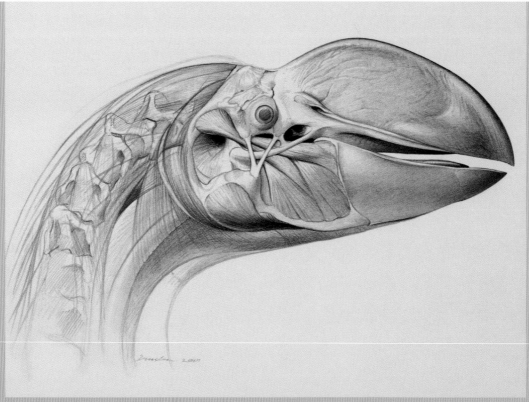

Figure 9.10 '*Dromornis stirtoni* – cranial muscle reconstruction' (2001). Graphite on paper, 38 x 52 cm, Queen Victoria Museum & Art Gallery collection, Launceston, Australia. The skulls reached up to 50 cm in length (P. Trusler).

Dromornis was big: it had a head the size of a horse's head! Furthermore, these birds had been described in the popular press as 'the demon ducks of doom', and the scientific debate about their dietary preferences had been seriously skewed in the mainstream imagination. This was an interesting point to me, because it highlighted the roles that palaeo-illustrators can inadvertently play in the popularisation of science. Amazing as these birds must have been, the story unfolding in the *Magnificent Mihirungs* was, although intriguing, not so emotionally charged.

I eventually centred my attention on the head of this bird, deciding that the resultant painting would be produced close to life size. This meant that I could confine the reconstruction to a more detailed study concentrated on the skull, and leave the extensive variety of illustrative work that Peter Murray had produced for presentation of the text. The stature of the birds could be accommodated with reference to the reconstructed landscape. I could show the massively-billed bird, leaning forward from the title, with sufficient room left over to incorporate the essential features of the landscape. The width of the painting's design could also wrap around the dust jacket or be spread over double pages if needed.

Figure 9.11 '*Dromornis stirtoni* – life reconstruction of head profile' (2001). Graphite on paper, 38 x 52 cm, Queen Victoria Museum & Art Gallery Collection, Launceston, Australia (P. Trusler).

The fundamental process for the *Dromornis* reconstruction was the same as for my other reconstructions. Perspective drawings of skull and neck anatomy were done in parallel with modelling-clay reconstructions of muscle systems. The sketches and photos of plants that I had taken in the field, as well as ones of magpie geese and screamers from captive waterfowl collections, were employed for comparative purposes. This informed both my depiction of the plant-processing capabilities of the bird, and the colours given to the sheath of the bill and the cloak of feathers. Despite the presence of fossils representing many typical groups of modern Australian plant

Figure 9.12 'Magpie Goose' (2002). Graphite on paper, 35 x 25 cm, private collection (P. Trusler) [above]. 'Black-necked screamer' (2002). Graphite on paper, 35 x 25 cm, private collection (P. Trusler) [opposite left]. 'Horned screamer' (2002). Graphite on paper, 35 x 25 cm, private collection (P. Trusler) [opposite right]. The primitive waterfowl that still survive in Australia (Magpie Goose) and South America (screamers) were important comparisons in the quest to understand the giant mihirungs. Adult Magpie Geese can reach up to about 1 metre in height, with a wingspan of 1.5 metres. Screamers can also reach 1 metre in height.

genera, at the time *Dromornis* was alive the structure of the habitat would have been very different to what we observe in those regions today. I hoped that my efforts would be visually apparent and act as an enticement to delve further into the pages of the *Magnificent Mihirungs*, a lengthy book with a wealth of anatomical, functional and family tree information. Along with my scientific companions, I also hoped that this work would place the mihirungs firmly in the ecological niches strongly suggested by their now well-known anatomy. Duck relatives they may be, but definitely not the demon carnivores of doom!

(Following pages) **Figure 9.13** '*Dromornis stirtoni*' (2002): reconstruction of Central Australian Miocene. Oil on linen, 71 x 96.5 cm, Monash Science Centre collection (P. Trusler). The final scene featuring *Dromornis* was produced as cover art for *Magnificent Mihirungs* (2003) and designed to represent a synthesis of the anatomical and palaeo-ecological theses presented by Peter Murray and Pat in this book.

10
Where the Wind Bites: Patagonia

Figure 10.1 '*Tehuelchesaurus benitezii*' (1998): reconstructed scene of the South American Jurassic. Watercolour, gouache and pastel on paper, 69 x 50 cm, collection of the Queen Victoria Museum and Art Gallery, Launceston, Australia (P. Trusler). Reconstructed scene featuring a sauropod dinosaur moments before being shrouded in deadly volcanic ash.

The Scientist: Pat

I've never been one to be bothered by motion, but the closest I ever came to being airsick was in a plane – sitting on the ground – in the Esquel Airport, western Patagonia. That feeling was preceded by our small plane touching down about 20 metres to the left of the airstrip! We had missed the runway because the wind was so strong (around 100km/hr) that it had forced the plane, if it were to land, slightly off course.

Wind … that's Patagonia. This is the land of incredible winds – a place 'where the wind bites'. There were days when my crew and I had been prospecting near the little settlement of Paso de Indios when the wind was so fierce that it blew rocks in my face – and those rocks were not so small!

Patagonia is many things: a wild place, one of the last wild frontiers on Earth. Tom and I, and our children and crew, went to Patagonia in our search for the bones and teeth of long-extinct animals, just as had Charles Darwin more than 100 years before. Darwin was a young man when he trod the vast deserts of this far southern land, but he found bones much younger than the ones we were after – his were only a few million years old and ours more than 65 million.

My crew and I were on the trail of dinosaurs of a similar age to those we had previously found in Australia. Tom, though willing to work with dinosaur bones, much preferred Mesozoic mammals. We had flown to Patagonia – part way on Qantas, the airline for which we had named a little plant-eating ornithopod, *Qantassaurus intrepidus*, recovered from the hard sandstones and silts that were exposed along the southeastern coast of Australia. This seemed an auspicious beginning. Maybe luck would be with us.

The dinosaur-bearing sediments that Tom, I and our crews had worked in Australia ranged in age from around 105 to 120 million years old. Similar-aged rocks, and some slightly older, also were exposed all over Chubut Province in Patagonia – some even Jurassic, the time period popularised by Spielberg's film.

Why search Patagonia and not some other place? Our decision to do this was in part careful calculation, but also more due to serendipity. In 1993 Tom and I were invited to attend a meeting to be held in the town of Trelew, initially settled in the 19th century by Welsh immigrants. Trelew lies well south in Patagonia, and is a place where one can still get a meal sporting the entire body of a cow or sheep, guts and all, along with a proper Welsh afternoon tea, ordered from a Spanish menu.

In 1994, the Museo Paleontologico Egidio Feruglio (MEF) held a meeting, the First International Symposium on Gondwana Dinosaurs. Ruben Cuñeo was the director of the museum, and he sent an invitation for Tom and me to attend. We managed to piece together funds for the airfares and began the long trek to get to Trelew – Melbourne, Sydney, Tahiti, Easter Island, Santiago, Buenos Aires and finally Trelew.

This was in April. Discussions continued with Ruben well beyond the meetings, held in one of the local hotels as the Museum was so small at the time there was no meeting space. Discussions continued while numerous drinks of strong tea (*maté*, drunk through *bombillas*, or metal straws) congenially passed around among all in attendance. After that came numerous incredibly delicious barbeques involving the entirety of sheep and cows – or, for that matter, anything else that runs about on legs, even armadillos! Unlike the Australian 'barbie', the *asado* included not only meat but also liver, gonads, brains, heart and so on – the lot.

From these discussions came an offer from Ruben for us to return and search for the elusive Mesozoic mammals and birds and anything else we could find. We were fascinated by the prospect and took Ruben up on his offer. We returned later in the year and began several years of exploration in western Chubut Province, centred on the Chubut River north of Paso de Indios where dinosaur bones had already been located by the Museo Paleontologico crew – Raul Vacca, Pablo Puerta, Eduardo (Dudu) Gomez, Daniel Lanus, a team of young volunteers and Ruben himself.

Figure 10.2 'Pat's prize locality' (2001): field sketch of fossil site from Chubut Province, Argentina. Watercolour on paper, 24 x 35 cm, private collection (P. Trusler). Los Altares landscape of Argentina; a wind-scoured terrain of Cretaceous sediments.

This exploration had marked relevance to our work in southwestern Victoria, at Dinosaur Cove, for when the animals whose fossils we were seeking were alive, South America and Australia were connected via Antarctica. These were the last remaining fragments of Gondwana and maintained their connection until as little as 100 million, maybe even 55 million years ago. So, some of the biota that we had from Australia could well have been shared with these other southern megacontinental bits.

Our work began in the far west of Chubut Province along the Chubut River, which, a bit like the Nile, flows across a vast arid plain, giving the impression of a narrow, verdant snake in an otherwise dun brown landscape. The Chubut has eroded into this landscape, which had been on the rise for millions of years as one part of the Earth's crust, the Pacific Plate, dived under the western part of South America along the Chilean Trench. Both the river-cut canyons and the ridges in this tormented landscape had yielded fossils, so back we trekked to these locales, to broaden the search and look for smaller things than the giants already known.

Our first stop was the Fernandez Estancia, approached by crossing the Chubut in a leaking rowboat, with one of the passengers providing the energy to move by pulling himself hand-over-hand along an overhead rope strung from one bank to the other. Awaiting us was the hospitable Fernandez family, who had run sheep on this land for years. That evening we had a full-on *asado* in the barn of the estancia and told tales to each other of our lives – in 'Spanglish', a combination of bits and pieces of both languages. I clearly remember Señor Fernandez, the same age as my own father, frustratedly saying to Maria Copello and Raul Vacca, who were translating for me, 'I wish I could speak directly to you – I have so much to tell you!'

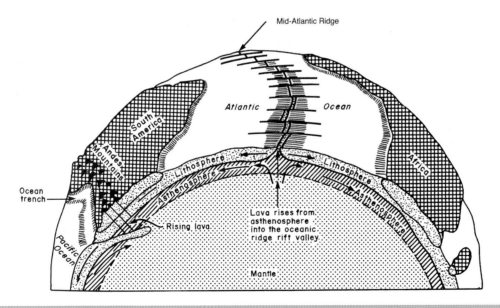

Figure 10.3 How the Earth works – cross-section of the planet illustrating plate tectonic theory (*The Fossil Book*).

Figure 10.4 Bones of *Tehuelchesaurus benitezii* in the ground on the Fernandez family's estancia along the Chubut River in western Patagonia (photo by P. Vickers-Rich).

The next morning, we all climbed from the flats where the Fernandez Estancia was built to the top of one of the ridges, where a full skeleton of a sauropod dinosaur had been discovered by the MEF crew. The wind was already blowing strongly, literally whipping up a dust storm. The next few days, as the multinational crew hacked out the hard sediments from around the skeleton, the wind laden with dust forced us all to wrap up our faces with scarves to keep our lungs cleared. It would take many further weeks after our crew left for the MEF technicians and volunteers to wrest this skeleton from its rocky grave and transport it to the museum in Trelew.

When we returned to Australia after our first field season, the bones of this gigantic dinosaur still lay encased in the rock. Two field seasons later this new dinosaur lay beautifully prepared in an alcove on the floor of the small Museo Paleontologico Egidio Feruglio, ready to be studied and already on show for the local public and the masses of tourists who flooded into this small town as a part of their cruises along eastern Patagonia to see the 'wolves of the sea', the killer whales, and the massive penguin colonies along this faraway coast.

The Fernandez Estancia dinosaur needed description, and this became our task, in between carrying out field work, on our next visit in 1997. My job, for the most part, was to photograph the specimen with the help of my teenage daughter, Leaellyn. To accomplish this, we had to move carefully among the bones strewn out on public display: she and I became part of the display for days, and answered many tourist enquiries. The bones needed proper lighting, and in some cases stereo photographs had to be made so that, in the research paper, readers could actually see the bones in 3D. Tom and I then had to describe these bones and work out if we had something new or just another specimen of a species already known. By using the literature in

Figure 10.5 Bones of *Tehuelchesaurus benitezii* laid on the floor of the old Museum Paleontologico Egidio Feruglio in Trelew, in much the way they were found in western Chubut Province of Patagonia. The length of the vertebral centra varies from 20–25 cm (photo by P. Vickers-Rich). Pat and Tom commenced their descriptions from this array.

the MEF library and papers that we had brought from Australia, together with useful suggestions sent to us via emails from the generous doyen of sauropod research, Dr John MacIntosh, we worked out fairly soon that the Patagonian skeleton belonged to a new species.

Then, began the very detailed work of comparing the Fernandez Estancia specimen, bone by bone, with all other known sauropods and showing how it differed. We wrote what is called a 'diagnosis', which is a point-by-point listing of the unique characteristics that made this dinosaur a separate species. In order to finalise this research paper, however, further work was needed. Some of the papers describing sauropods were excellent, with detailed descriptions and high-quality photographs or drawings. Others were not, in part due to specimens being incomplete. What we needed to do was visit museums around the world where either one or both of us could examine each of the named sauropods and compare the new Patagonian specimen directly – using our high-quality photographs and the detailed descriptions based on our own examination of the specimen.

Over the next couple of years, we managed to visit those collections that had full skeletons, as far away as Sichuan in China. After detailed comparisons, we felt sure that the Fernandez Estancia specimen was a new species, and we were able to complete the paper with input from many of our colleagues, whose names appear on the author list.

The final question was what name to give this form – and one of our field colleagues, Daniel Rohas Lanus, suggested that it would be appropriate to use the name of the local Indian tribe that had once occupied this area. And so *Tehuelchesaurus* was written into the dinosaur records. Our sauropod's specific name, *benitezii*, honoured Aldino Benitez, who had discovered its disarticulated bones. Our description of this new species formed the basis of a paper published as part of the

Figure 10.6 Fossil araucarian branches preserved in the lake sediments at Turtle Town, to the south of the locale where *Tehuelchesaurus* was found. This sort of vegetation was common during the Jurassic and Cretaceous in this region and forests of such trees still exist in the wetter mountains to the west of the country, which today is generally much drier (photo by P. Vickers-Rich).

Figure 10.7 Skin impression of *Tehuelchesaurus* (photo by S. Morton).

2nd International Symposium on Gondwana Dinosaurs, held in Tokyo in 1998. Peter was engaged to paint a reconstruction of this new dinosaur, and that reconstruction graced the cover of the program for the Gondwana dinosaur symposium. It was also used as part of the newly discovered material highlighted in the revision of *Wildlife of Gondwana*, published in 1999 by Indiana University Press.

Peter's reconstruction of *Tehuelchesaurus* was typical of his art. It was as close to reality as he and we could get. The illustration depicted the middle of the dinosaur. Quite rightly. *Tehuelchesaurus*'s head is only viewed from behind, and the last half of the tail is out of the illustration. Why? Because those parts of the skeleton are missing, never found in the Fernandez specimen, and so Peter had nothing to work with! In the reconstruction, *Tehuelchesaurus* is gazing at an approaching cloud – which happens to be a cloud of ash explosively ejected into the air by a nearby, menacing volcano. The ash was soon to engulf this hapless reptile and preserve it in exquisite detail until discovered and excavated more than 170 million years later by the palaeontologists from the MEF. Associated with the bones of *Tehuelchesaurus* were impressions left by the skin of this massive reptile – a mosaic of minute scales. So, Peter did not have to guess at what the outside of this long-lost giant looked like. Even the surrounding vegetation is depicted as accurately as possible, based on the fossil flora associated with the Fernandez specimen. Ruben Cuñeo, a palaeobotanist, had himself described the very cones, leaves and other plant debris that clearly represented the flora of the time and likely formed part of the food *Tehuelchesarus* enjoyed.

Association with the MEF continues to the present day, with most of the field work we were involved in there lasting for a decade. I made sure that, for background, Peter was able to travel there, along with a previous graduate student of mine, Andrew Constantine. Making two trips to Argentina, Andrew was able to reconstruct the ancient environments in which this massive dinosaur lived by mapping and interpreting the sediments that had encased the *Tehuelchesaurus*. The final trip included Peter, Tom, Andrew and me. Because I unfortunately had heart failure on this trip, after an expedition down the Amazon as a lecturer on a National Geographic ecotourist trip, I sent the rest of the team to the field. In the meanwhile I recovered sufficiently in Trelew, under the care of brilliant cardiologist Dr Alejandro Sarries, who had, fortunately for me, settled in this country town to escape the distractions of the big city. He, and my good friend Maria Capello, kept a close eye on me while I recovered. Peter and the crew were able to visit not only the rocks from which *Tehuelchesaurus* had been recovered but also the living araucarian forests in western Chubut, the closest approximation of the ancient forests that our sauropod had inhabited.

On one of our earlier field excursions, we visited the La Colonia Formation to the northwest of Trelew – dark sediments that appear to have been deposited under anoxic estuarine conditions. We were particularly interested in the possibility of finding small vertebrates here, as our close friend Rosendo Pascual and his team had already found them to be present. Our hope was to find some mammalian remains, and on one of our forays one of my graduate students, Corrie Williams, found what Tom and I identified as the premolar of a multituberculate, very similar to the North American *Mesodma* – an extinct order of mammals that resembled rodents but were not directly related to them. I remembered the morphology of this particular primitive mammal, for

Figure 10.8 'Southern conifers and condors – field study' (2001): *Fitzroya cupressoides*, Alerces National Park, Argentina. Graphite on paper, 38.5 x 28.5 cm, collection of the artist (P. Trusler). The magnificent remnant of the old growth forests in western Argentina. These field studies show samples of the same plant types as depicted in the South American Jurassic reconstruction painting (figure 10.1).

Figure 10.9 '*Fitzroya cupressoides* – field study 1' (2001): Alerces National Park, Argentina. Graphite on paper, 38.5 x 28.5 cm, collection of the artist (P. Trusler). Sketch of an old wind-damaged giant in western Argentina.

Figure 10.10 '*Fitzroya cupressoides* – field study 2' (2001): Alerces National Park, Argentina. Graphite on paper, 38.5 x 28.5 cm, collection of the artist (P. Trusler). Botanical field notes, western Argentina.

Figure 10.11 'Araucaria' (2001): study of immature male cones and leaf whorls of monkey puzzle tree, *Araucaria araucana*; plantation specimen, Funtulafquen, Alerces National Park, Argentina. Watercolour on paper, 34 x 26 cm, private collection (P. Trusler).

I had found another specimen on my wedding day (3 September 1966) at Bushy-tail Blowout, northeast of Lance Creek in Wyoming. Little did we know that years later another multituberculate would be discovered at one of our sites in Australia; this was an important find because most of these primitive mammals are known from the northern continents. Peter would, of course, then be engaged to reconstruct that tooth and it would in part be named after Corrie (*see* Chapter 12).

On my next trip to Argentina following completion of the *Tehuelchesaurus* reconstruction, I brought a high-quality print of the painting and asked the brother of one of the MEF crew, Maria Copello, to make a beautiful frame for it. I then flew to Trelew to begin my work there, and presented the framed print to Director Ruben Cuñeo, and it hangs in his office today.

From all this collaboration came two projects that I had not ever imagined at the beginning of our work in Chubut. One was for me to be part of the planning and completion of a brand new museum building in Trelew for the Museo Paleontologico Egidio Feruglio. Ruben asked me to be involved in the design and completion of this, and I was delighted. The second project was the placing of a major exhibition on the history of life on Gondwana in the main natural history museum in Buenos Aires, the Museo Ciencias Bernardino Rividavia. Our work with the MEF certainly facilitated this, and the funding generated by this latter exhibition greatly assisted further research programs in Chubut. Besides that, the exhibition was profoundly successful in Buenos Aires. I remember receiving an email from Eduardo Romeo, the director, which asked: 'How do I handle success?' On the day the exhibition opened, more than 2000 people were waiting at the door – compared to the normal attendance on a weekday, which would have averaged around 100. This was in part due to the draw of the dinosaurs on display, but was greatly facilitated by the advertising campaign provided by Qantas Airways, as well as the enthusiasm of the then Australian ambassador and her staff in Buenos Aires. *Qantassaurus* was one of the stars of the show, and *Qantassaurus* was most certainly Australian! Success was also nurtured by our friendship with researchers in the Argentine science academy, the CONICET, an organisation that supports a wide range of scientific projects and staff. This was a case of researchers, private industry and governments all working to a common goal.

Ironically, simultaneous to our exhibition launch, the dinosaur cartoon movie *The Land Before Time,* straight out of Hollywood, burst on the scene in Buenos Aires. The exhibition and the movie most certainly cross-promoted. On the opening night, the Australian ambassador and I gave short talks at the movie's premiere in one of the major cinemas in downtown BA. My talk was illustrated by slides on the big screen, most of which were palaeo-reconstruction art by Peter. Then we all sat back and enjoyed the show. As I watched this whole scene, it brought back memories of similar events in Australia in 1993, with the simultaneous launch of the *Great Russian Dinosaurs* exhibition, Steven Spielberg's *Jurassic Park* and Australia Post's Australian dinosaur stamps! The public reaction was overwhelming and the attendance at both movie and exhibition spectacular. Hollywood certainly has a place in promotion of science – but one must attend to the detail!

The Artist: Peter

One of the most significant features of any deliberations in palaeontology is regard for matters that are not known. The record is, by its very nature, fragmentary, and our view of moments in time is based on a multitude of opportunistic 'windows'. These windows are seldom fully transparent; the view is as if seen through filters that provide only a little of the information that could have been in the picture. The ultimate task is to painstakingly assemble the ever-expanding album of information from every individual glimpse thus obtained. It is this that leads to our understanding of the past: it begins to build an idea of the biota, an understanding of the environment and a comprehension of the pattern of processes that have interacted to produce the changes on earth throughout time.

Another key part of that process of information gathering from each of these windows, and of understanding any patterns that emerge, is data and hypotheses that can then be tested against other input gathered from new windows.

My illustrative presentation of a partial sauropod dinosaur specimen from the Jurassic of Argentina encapsulates such a process. Here I was presented with a well-preserved, articulated specimen about which a great deal was known: the environment in which it lived and the circumstances in which it died. The detail of the vertebral column alone was sufficient to indicate that this creature was new to science, but its exact identity remained a mystery because the skeleton had no associated head. This situation was not unfamiliar to me, and is certainly a scenario familiar to palaeontologists. My wildlife watching and beach-combing experiences have informed me that the relatively delicate or complex structures, such as the heads of animals, are often broken or completely detached from corpses either in the process of decay or through scavenging. Finding the torso or only the most robust bones is the more likely scenario. Palaeontologists have devoted entire branches of study to specialised parts of animal anatomy, such as teeth, simply because they are the only parts of the skull that commonly survive. For instance, Tom's years of work looking for Mesozoic mammals in Australia have revealed many specimens, virtually all isolated teeth and fragments of lower jaws. Detailed studies and meaningful comparisons can still be made through the collection of a series of similar and commonly preserved skeletal elements, whether they be teeth, thigh bones or the central portions of vertebrae. The studies of invertebrate palaeontology are often based exclusively on the external hard parts of animals, such as shells, cephalopod beaks and spicules. I have often illustrated these for the record and for the purposes of comparison to better-known fossils or living animals.

The discovery of the fossilised remains of such a massive animal from which the head was missing – even though such a head would have been relatively large – therefore seemed unremarkable. Although its incompleteness was a little disappointing to me on one hand, it was still important to attempt the communication of the story around its discovery.

The reconstruction of images of the past always has to bridge these gaps. The story, as best we understand it, often needs to be told in part, with some chapters missing. Each discovery and each retelling of the story, and those related to it, may add another chapter or even bring a new paradigm, a new set of rules, which can fundamentally change our perceptions of the past. The science, and its illustration, both evolve. The vital point is that we must strive to tell what we know, and be conservative with that information, remaining cognisant that there are things we are yet to ascertain. If I can do this in such a way that allows me to incorporate as few subjective inclusions in the artwork as possible, the strength of the science will be supported. I know that, in time, my art might become outdated with advances in research, but it remains important that the illustrative function serves what is known at the time of rendering, with fidelity and for as long as possible. It is equally important that the art does not of itself serve to limit the potential for enquiry and further discovery. It needs to stimulate the mind and assist in posing questions that require answers or inspiring the testing of present ideas. To this end, the gaps in our knowledge actually need to be made apparent.

How could I reliably depict a dinosaur without a head – as well as show this once-living animal in the habitat of its time and the events that led to its partial preservation? The body indicated that the head was likely to be different from others of its group, but how so? This could not be reliably depicted, even though much of the rest of the sauropod could, except for the tip of its tail. In this instance the ash had even moulded the detailed record of sections of the scaly skin of *Tehuelchesaurus*. It was if I had the fingerprints of this beast, but no record of its face. I had dealt with similar issues of incomplete data in both minor and fundamental ways before, and so positioning the tail of this sauropod to trail out of the composition took little imagination. Positioning unknown portions of a given animal in the undergrowth, representing environments in part or by reflection, or simply isolating a subject or fragment of a specimen against a plain background – I had done all this before. In more academic illustration work where I had depicted entire subjects, I had in the past represented those features of the specimens that required interpretation with some contrasting or delineating technique, clearly informing the viewer where the facts ended and the estimates began. There is an art to the decisions that need to be taken in presenting information and in dealing with the uncertainties and gaps in understanding. The ways that might best be used are, of course, very dependent on the purpose of the illustration being produced.

In the case of poor *Tehuelchesaurus*, the cloud of volcanic ash that descended on its forest home was depicted advancing towards this dinosaur, and the hypothetical witness can envision the unfolding event. By placing the viewer in the position of witnessing the event as *Tehuelchesaurus* did, I hoped to invoke enquiry and at the same time depict all the known gargantuan features of the beast from behind and not need to depict the specific, and as yet unknown, details of its head. In this way the identity of this sauropod, intimately known in some respects yet not in others, did not detract from the communication of information about the wider environment and events leading to its death – preserved some 170 million years ago near an active, and deadly, volcanic field in the far south of Patagonia.

Figure 10.12 'Cerro los Chivos' (2001): field sketch from Chubut Province, Argentina. Watercolour on paper, 23 x 35 cm, private collection (P. Trusler). The eroded landscape revealing the volcanic necks that have enabled the dating of the sediments containing *Tehuelchesaurus*.

Figure 10.13 View from the site where *Teheulchesaurus* was found with the River Chubut in the background (photo by P. Vickers-Rich).

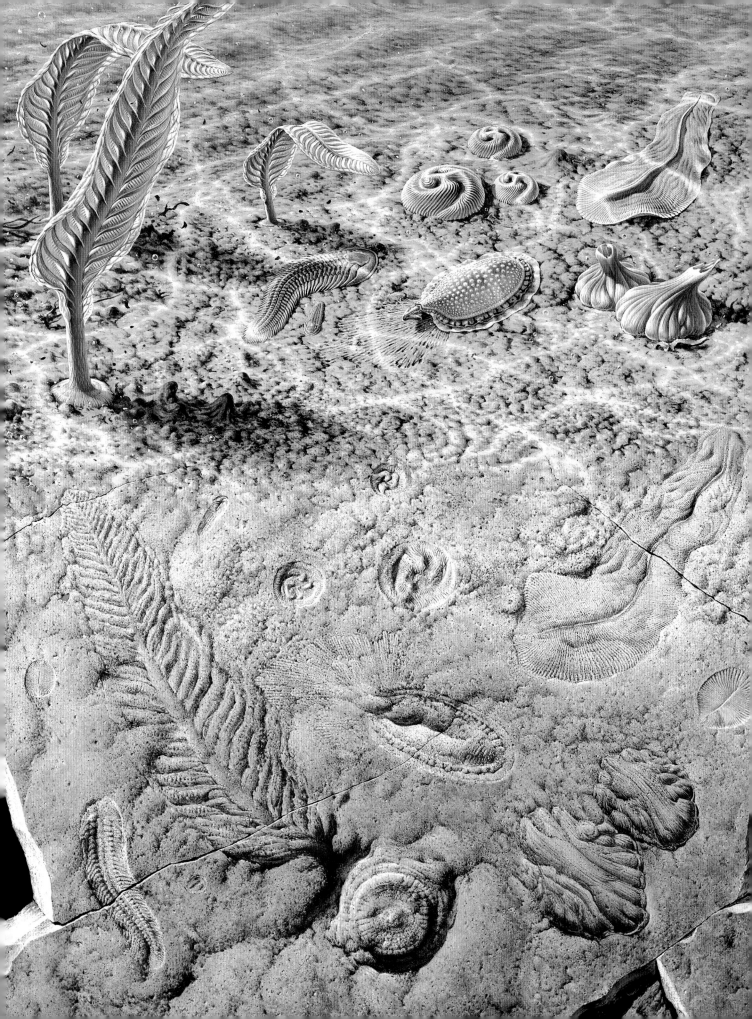

11

The Rise of Animals: Back to the Precambrian

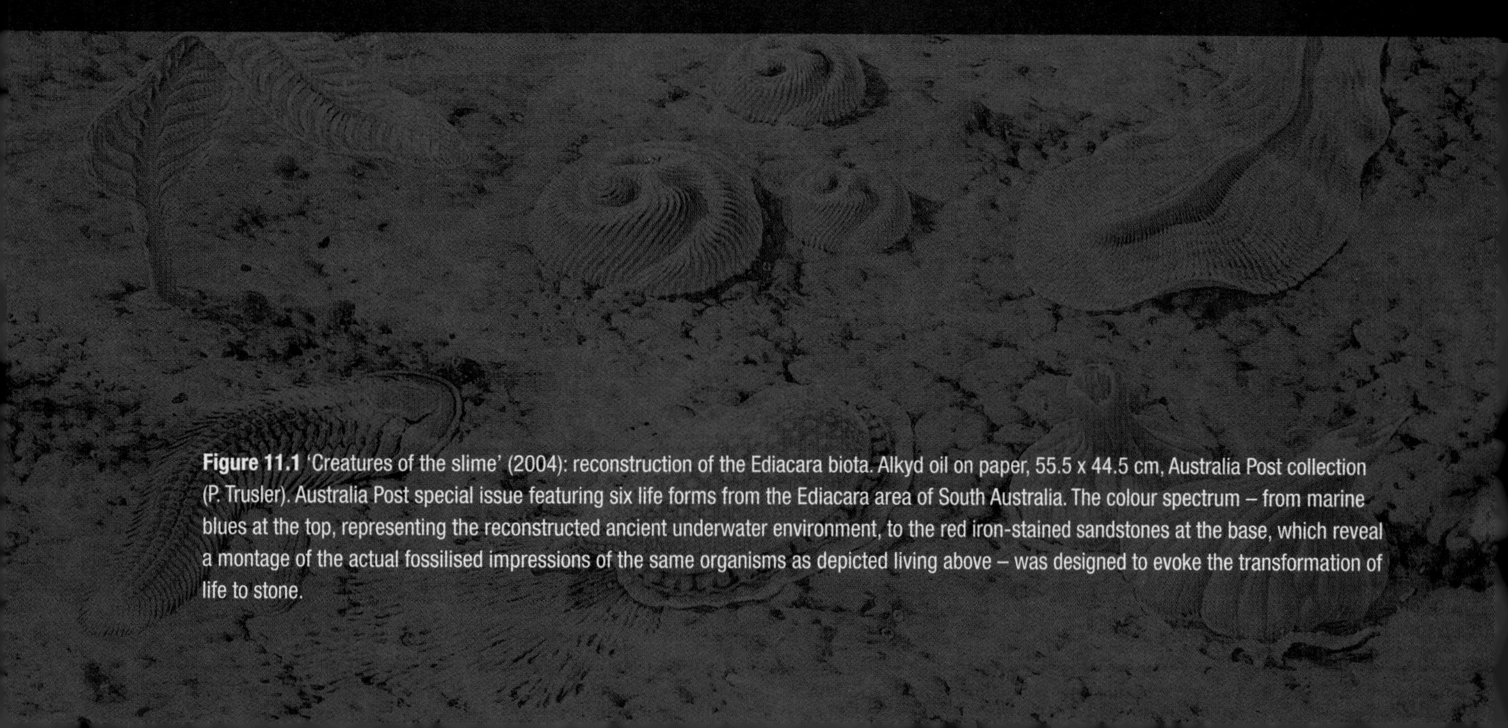

Figure 11.1 'Creatures of the slime' (2004): reconstruction of the Ediacara biota. Alkyd oil on paper, 55.5 x 44.5 cm, Australia Post collection (P. Trusler). Australia Post special issue featuring six life forms from the Ediacara area of South Australia. The colour spectrum – from marine blues at the top, representing the reconstructed ancient underwater environment, to the red iron-stained sandstones at the base, which reveal a montage of the actual fossilised impressions of the same organisms as depicted living above – was designed to evoke the transformation of life to stone.

The Scientist: Pat

In 1993, I made my second trip behind the Iron Curtain to Moscow – although the curtain was certainly neither iron nor difficult to pass through. It was relatively easy to gain a visa, though the wait to go through customs and immigration at the main airport, Sheremetyevo, to the northeast of the city, did take a few hours. This trip was with a small delegation from Museum Victoria, Monash University and the Queen Victoria Museum and Art Gallery in Launceston, Tasmania (QVMAG). The group included Tom, David Smith, Frank Coffa, Chris Tassell and me. Tom and I (and, to a degree, Chris) were the research scientists; Chris and David were museum administrators and also scientists, Chris being the director of the QVMAG. Frank was our photographer, head of photography at Museum Victoria and a long-time companion of Tom and me on our expeditions, photographing specimens, field areas, modern fauna and flora – just about anything around.

Our task on this visit to Moscow was to negotiate a contract with Alexei Rozanov, the director of the Paleontological Institute (PIN), and his staff to bring a collection of Russian dinosaurs and mammal-like reptiles to Australia. Not only did we have to negotiate the contract, but we also had to select the specimens and photograph each, so that we could have the exhibition and a showy exhibition catalogue published by August of 1993. Given that our visit was in late May and early June of that year, we had a real task on our hands, and one that I would never take on again in such a short time frame.

Our negotiations were relatively short and agreeable, and the promise by Qantas Airlines to transport the goods, nearly a bellyful of freight for a 747, came through at the eleventh hour. So, much of our week and a bit in Moscow was spent foraging in the collections, determining the specimens we wished to include and getting them photographed by Frank. My task, besides being head negotiator, along with Chris, was to gather all the information that I would need to quickly write and design the catalogue. Peter's work, which had graced *The Fossil Book* and was being prepared for a near-simultaneous launch of the 1993 Australian Dinosaurs stamp issue, was used in the exhibition catalogue, for there was no time to commission new work. We had two months to pull all this together and get the exhibition in place. The catalogue was produced and handed over, all printed and bound by our miracle-working publisher Derrick Stone, on the eve of the day Gareth Evans, the then Australian Minister for Foreign Affairs, was to launch the *Great Russian Dinosaurs* exhibition!

One of the Paleontological Institute staff sitting beside Chris and me at the contract negotiating table was Mikhail (Misha) Fedonkin. Misha casually mentioned to me, in between negotiations, that he had been working up along the White Sea coast in northern Russia and had found some amazing fossils with exquisite preservation. They were much, much older than our exhibition specimens of dinosaurs and mammal-like reptiles, and much, much more interesting than dinosaurs! He offered to take our group to the part of the PIN where the big slabs of claystone lay spread out on old wooden cabinets. I was truly amazed at these very strange looking macrofossils; I had never seen anything like them. And the detail – so good that it looked like what you might expect if you pressed a just-dead modern species into soft mud and could see every little structure. When Misha told me how old he thought they were, I was doubly amazed – more than 550 million years old. And the claystone, he said, even though hard now that it had dried, you could stick a knife into when it was first exposed up on the cliffs of the White Sea.

Figure 11.2 Mikhail (Misha) Fedonkin (left), Pat and Dmitri Grazhdankin along the White Sea coast, northern Russia (photo by N. Hunt).

That was 1993. The *Great Russian Dinosaurs* exhibition did come to Australia. I shepherded it around Australia and the United States to a number of museums until nearly the end of 1997. Millions of people came to see it, it generated millions of dollars, which helped the PIN to survive the hard years of the new Russia, and I was able to run the fledgling Monash Science Centre (MSC) and encourage the use of our share of the takings to generate further funds for the building of the permanent quarters of the MSC (Building 74 on the Monash University Clayton campus), which was opened in 2002.

The lead up to the launch of this building was frantic. Not only did the building itself have to be completed, it also had to be entirely outfitted – all of the new exhibition space properly equipped, the decking laid out of the café that was to serve the first coffees to the guests. The speeches had to be honed, and the first major exhibition, *Wildlife of Gondwana*, had to be put in place. This was no small task, as it included several large skeletons, all the information panels, and all the education materials. And, of course, to top this off, all of my staff and our possessions had to be moved from our old quarters on the other side of the university and put in place in their new home.

By the time we got ready to launch, I was a wreck, but the show had to go on. I was the MC, and to this day I will never forgive myself for forgetting to ask the dean of science, Rob Norris, to give his speech. He, along with our deputy vice chancellor of research, Peter Darvall, and our vice chancellor, Mal Logan, had been some of our key supporters, the latter two critical in the realisation of our new building. (Fortunately, I remembered these fundamental backers, Peter Darvall and our VC, along with John Trembath, in other ways – naming JP's, our coffee shop after John and Peter, and acknowledging Mal on the brass plaque in the entry foyer of the science centre. Yoshikazo Hasegawa had flown all the way from Japan to be there for the opening – his part in bringing about the *Great Russian Dinosaurs* tour was also critical to the final realisation of the science centre.) Forgetting the dean of science in the speech roster was truly a sign of my exhaustion. Fortunately, I was gently looked after by Peter Frampton from Qantas Freight and Kay Hamilton, our main public relations advocate, after the ceremonies and taken for a quiet dinner for my core crew at the Science Centre – Peter Darvall and my crew had been so critical in the gathering of funding to make this venture succeed. I am thoroughly convinced that individuals *can* make things happen. Often big organisations can only manage this if one or two strong individuals are willing to roll up their sleeves and strive to put programs into action. The Monash Science Centre would not exist had it not been for a few such individuals.

The building was duly launched, and then our scientists-in-residence arrived. First came Alexei Rozanov, Director of the Paleontological Institute, who had been so instrumental in the success of the *Great Russian Dinosaurs* exhibition and was, therefore, a person to be thanked for the existence of our building. We discussed his work on the Precambrian – his main research being directed at microbes. A little later the second of this research pair, Misha Fedonkin, arrived for a few months' stay. He lived with Tom and me during this time. We began discussing, both at work and over our evening meals at home, those strange fossils that Tom and I had seen in Moscow in 1993. I had become completely dazzled and very curious about these ancient animals – if animals they were – and over the next few months began to be drawn away from my previous research. In fact, I wanted to do something different, work in a new area and work on my own.

By the end of 2002, I was hooked on the Precambrian and increasingly curious about the conditions that had nurtured and spurred on the development of the first animals. I read everything I could about this and had a lively correspondence with Misha, which has continued for years. Misha introduced me to many of his long-time colleagues, who have since become both my friends and my research colleagues. As I began to move into this area, I felt the need to become acquainted with the fossils from the many different areas of the world where the oldest remains of animals were known – and one of the first places I travelled was back to Moscow to look in detail at those enigmatic forms that Misha had shown us in 1993. I also wanted to go in the field, to see where they had been collected.

Early on, I had begun to collate all of the new information I was gathering and then thought to myself, why not write a book based on this compendium, since there was nothing published of this kind. My particular interest in this was to put together an atlas of all the species of these early animals that had been described. I was already doing this for my own edification, so why not publish this work? At first, I was discouraged from doing this, because many of my new colleagues told me it had already been done; however, they did not fully understand what I had in mind. What was out there was a few pages in a great book – organised by two giants of Precambrian research, Bill Schopf and Charles Klein – called *The Precambrian Biosphere: A Multidisciplinary Study*, published by Cambridge University Press in 1992. In that book, Bruce Runnegar and Mikhail Fedonkin had several pages listing the different taxa and some information on each. But, because I have a visual mind, what I wanted to create was a compilation containing illustrations of each species, with a much expanded database for each one of them.

And so I began this compilation, as well as gathering as much information as I could from my increasing group of Precambrian colleagues. I had in mind a book that would explore not only the different organisms that made up this entourage of early animals, but the places they came from – the White Sea of Russia, the deserts of Namibia and South Australia, the blustery coasts of Newfoundland, and several other lesser-known places such as sub-Himalayan India, Siberia, the Ukraine, the cold mountains of Canada, even the Yangtze region of China, and more. I wanted to visit all of these places and document their settings for myself, along with experienced researchers. And, I also wanted to obtain images of as many of these places and fossil species as I could. I took along one or more of my photographers, Frank Coffa or Steve Morton (from Monash University) – by 2004, I had made one trip with Frank to Moscow to photograph many of the specimens. I was also given access to the material of other researchers and their photographs. And, of course, Peter travelled with us too. In addition to studying all this material, he needed to see these places and specimens personally, for this was yet another opportunity for his reconstruction art – it would play an important role in the success of any book I was to organise on this topic.

But really, it was not just my book: the idea was dreamed up by Misha Fedonkin and me while he was the scientist-in-residence at the Monash Science Centre. Beyond our efforts, Jim Gehling of the South Australian Museum, an expert on the Ediacaran fauna of the Flinders Ranges, was invited to contribute, as was Guy Narbonne of Queens University in Canada, whose work in Newfoundland and the Mackenzie Mountains in Canada's northwest was critical to any summary book. Kath Grey, a micropalaeontologist from the Geological Survey of Western Australia, took on the challenge of writing about the acritarchs, marine microfossils that lived alongside the metazoans – the animals that, as well as the reef-forming stromatolites, were abundant during the time animals were making their first appearances, and long before. Many other colleagues contributed as well: in particular, Soren Jensen from the Universidad de Extremadura in Spain; Maxim Leonov, a graduate student in Misha's Precambrian Lab (PIN) in Moscow; my research assistant, Patricia Komarower; our nitpicking proofreaders, Tom Rich and Mary Walters; and our tireless draftswoman, Draga Gelt. In fact, in the end it was Draga and I that provided the design work for Johns Hopkins University Press, which in 2007 published this 326-page full-colour book, *The Rise of Animals. Evolution and Diversification of the Kingdom Animalia*, with a foreword by science fiction icon Arthur C. Clarke. In other words, though I was the circus coordinator, there were many actors in the circus tent.

Figure 11.3 *Pteridinium* fossils from Farm Aar in southern Namibia, Africa: despite their plant-like form these appear to be animals, although scientific consensus has not been reached on this question. The head of a rock hammer is on the left for scale (photo by M. Fedonkin).

Figure 11.4 Charlie Hoffmann (left), chief geologist at the Namibian Geological Survey, and Mikhail Fedonkin, head of the Precambrian Laboratory at the Paleontological Institute in Moscow – teaming up with Pat and others from the PIN in Namibia (photo by P. Vickers-Rich).

My drive to visit as many of the centrepiece collections, and likewise the sites where they came from, was first put into action along the shores of, and up the rivers feeding into, the White Sea to the west and northeast of Peter the Great's mercantile town of Archangel'sk, far north of Moscow. Peter (Trusler, that is!), Tom and one of my previous students, Andrew Constantine, were also able to visit this remarkable fossil field. Peter gathered considerable information on the setting of these fossils, together with a serious crop of insect bites. Andrew was there to map the sediments and work on the environmental setting they reflected. Tom was along to learn about the unique manner of occurrence of these unusual fossils in outcrop, information always of use to someone exploring for new fossil fields. And there were many of the Russian researchers, including Misha Fedonkin, who had organised this expedition and had spent years and years gathering the metazoan treasures these sediments had entombed.

Just as I had noted in 1993, the sediments along this wild coast, more than 550-million years old, were unexpectedly soft and nearly undeformed. Originally deposited nearly horizontally on an offshore seabed, they had been only slightly tilted and had not been deeply buried. They were more like the Pleistocene marine sediments that I had worked with in the past – those less than 1 million years old. I had to continually remind myself that these were more than 500 times older!

One other result of my visits to the collections in the Paleontological Institute (PIN) in Moscow in 2003, and to the White Sea in that same year, was the planning for a long-term project which Misha Fedonkin, Jim Gehling and I put together. I submitted a proposal to embark on a five-year project under UNESCO's International Geological Correlation Program (IGCP). This project aimed to consolidate the available data on the first metazoans that appeared during what was then called the Vendian Period (now renamed the Ediacaran), and to find more. We were seeking out the factors that had nurtured the appearance of this new group of organisms – some of the oldest evidence of megascopic, multicelled life forms – and were equally interested in determining the cause of their demise. So, the title of our project, to become IGCP493, was *The Rise and Fall of the Vendian Biota* (www.geosci.monash.edu.au/precsite).

Figure 11.5 *Kimberella* is perhaps the oldest known mollusc. Its form is known from over 800 specimens from the White Sea coast and it also occurs in Australia. Records of its feeding scratches in the ancient biomats have been preserved in both locales. *Kimberella* is one of the few of these early metazoans that may have given rise to, or be related to, animals that took over the Earth at the beginning of the Cambrian. So many of the other Ediacarans seem to have left no descendants beyond this transition. The central shell with ornament reached up to 15 cm in length (photos by M. Fedonkin).

Figure 11.6 *Rangea* from the White Sea of northern Russia. Rangeomorphs are not common in this location, but belong to a group that had a long history in the late Precambrian, known from the oldest macro-metazoan–bearing sediments in Newfoundland to the youngest in Namibia. This specimen is 63 mm wide (photo by M. Fedonkin).

Figure 11.7 Jim Gehling (left), Mikhail Fedonkin (right) and Pat walking across early metazoan (Ediacaran) infested rocks in the Flinders Ranges of South Australia, one of four important locales on the globe that has yielded fossils of the oldest animals (photo by F. Coffa).

So began a series of projects that ranged from field trips for Peter and me – to the White Sea, Namibia, South Africa, Newfoundland, Japan and other places – to symposia in Prato (Italy), Kyoto (Japan) and Moscow. Several documentaries were also the offspring of this project, as were exhibitions and even another Australia Post stamp issue in 2005 (*Creatures of the Slime*). Peter was commissioned to produce many works, which ranged from popular, yet scientifically accurate, to detailed research drawings, and this art has been picked up by many other researchers, teachers and the media to this day. The art was funded both privately and again by Australia Post, and has truly had broad use. And, in order for Peter to have the best chance to ply his trade, I took on the responsibility for curation of some of the critical collections in Africa and Russia. More than $100,000 would be needed to underwrite those programs, and as time would allow, I made that my goal. Luckily, I succeeded – due to the generosity of individuals like Nathan Hunt (and his Nebraskan family) and the income from travelling exhibitions that I organised, as well as a variety of other sources that provided significant support to fund the building of specimen cabinets, purchase computers and underwrite the salaries of technical staff in several collections around the globe. Peter could then see the wealth of these early metazoans, some of which had been scattered in institutional collections for decades and were not easily accessible.

As Peter steadily worked on this Precambrian art, a number of researchers who had long been working on the Neoproterozoic metazoans realised that his art was not exactly what they were used to seeing in reconstructions: there was a serious attention to detail. And they were not quite used to dealing with an artist who questioned their interpretations. This was common practice between Peter, Tom and me. Many serious discussions had passed between us, and the outcome, the final art, was always an amalgam of our combined ideas. It became clear to our new colleagues that Peter was not just 'the artist' but was also a co-researcher, and as time went on his name began to appear on the author list, not just in the acknowledgements as the artist.

Peter and I were both 'new kids on the block' in the Neoproterozoic metazoan neighborhood – and, I have to say, we were welcomed and our observations taken seriously by this new mob. We have both cherished this acceptance by our new colleagues. To both of us, and Tom as well, this represents the true spirit of science.

On my first field expedition to the White Sea with Misha and his team, we travelled from Archangel'sk to the river Suz'ma off to the west in a Russian military helicopter, an MI8. The pilots of these leftovers from the days of the Soviet Union were exemplary, coping admirably with both the older aircraft and the local weather conditions. We were dropped off at a small village and travelled upriver from there on foot to some small outcrops along the eastern side of the river. On one of the first days we were there, I picked up a specimen that had some odd markings on it. When Misha saw it, he immediately said that it looked like another specimen he had found years earlier, a form with some similarity to yet another singular specimen, *Ausia*, from Namibia. *Ausia* had been found – perhaps by German soldiers stationed during the late 19th century in what was then German Southwest Africa – and eventually published on, decades later but before Misha's and my finds. The German huts – as well as the remnants of the much earlier Bushman habitations – can still be seen there, sentinels with the remains of some of the very primitive animals we were studying embedded in the slabs making up these simple dwellings.

Figure 11.8 *Parvancorina*, an Ediacaran metazoan and possible trilobite relative, from the Flinders Ranges, South Australia. The width is about 2.5 cm (photo by S. Morton).

Figure 11.9 Reconstruction of an Ediacaran fossil showing the burial event. Alkyd oil on paper, 34 x 26 cm, private collection (P. Trusler).

I asked Misha what *Ausia* was – and his reply was interesting. He said that the current assignment of *Ausia* was to a group that included forms thought to be distantly related to vertebrates – the tunicates, commonly referred to as sea-squirts. This group has a varied life history where the juveniles are mobile tadpole-like forms. The adults are sedentary, and when disturbed they contract and literally squirt out the water that is being filtered through their branchial baskets – their main food collector. The assignment of *Ausia* to this group had been based on the complex pattern imprinted in the fine-grained sandstone of the Nama Group rocks that enclosed *Ausia*; this pattern so closely resembled the branchial basket in living tunicate adults that there was really nothing else comparable in the living biota.

Peter, Misha and I then embarked on a detailed reconstruction of these two rare specimens, taking into account *Ausia*, and when all of us later visited Namibia Peter was able to examine that specimen in detail. We were all interested in the geological context of these specimens; besides the detailed sketches and photographs of the fossils themselves, we reconstructed the scenario that we proposed as the mode of preservation. This hypothesis was supported not only by our observations of the morphological detail of the specimens, but also by such information as clay plugs in some of the fenestrae (openings) in what we interpreted as the branchial basket.

Figure 11.10 Helicopters are an excellent mode of transport in the high Arctic – the pilots do a superb job of keeping their ageing machines in the air in very challenging conditions. We used them on the White Sea coast; the one shown here is dropping palaeontologists in the Siberian section, where other Ediacarans have been discovered (photo by M. Fedonkin).

Figure 11.11 '*Yorgia*' (2003): field study of Precambrian fossil from Russian expedition, June–July 2003. Watercolour and gouache on paper, 36.5 x 27.5 cm, private collection (P. Trusler). In the grey environs of the White Sea coast, northern Russia, the fossil *Yorgia* has been represented as a 'ghost' in the sky.

It appeared to us that the living individuals of the new form from Russia had been overwhelmed by a massive downslope avalanche of sediments, which included rip-up clay clasts and sand. As the turbid water entered the branchial basket of this bottom-dwelling animal, some of the clay clasts made their way into the branchial basket and lodged in a few of the openings as the water-soaked sediments were pushed through. These organisms were then completely buried in the sediment wedge, where they were compressed. The presence of these individuals was not noted again until the sediments were uplifted and eroded, and Misha and I came along and discovered their fossil impressions. That, at least, was the scenario that we proposed, based on our observations in the field and in the lab.

So, in the investigation of these possible ancient tunicates, Peter and I had to first work out, and then describe, what we were looking at: an internal impression, an external impression, or something else, based on what we knew about what we thought were the closest modern analogues. These were the research drawings. Then came the scenario reconstruction, a series of schematic sketches, and finally Peter produced an imaginary painting of what we thought the last moments of this possible tunicate from Russia might have looked like – an individual about to be engulfed by a terrifying (to us, but probably not the tunicate!) shallow marine downslope avalanche (*see* figure 11.9).

Our work on these early metazoans continues today, and has required Peter to employ a number of different art forms – both for the general public and for research papers that put forth images as hypotheses to consider. All of the art has been, as usual, clearly based on scientific data – be it in research papers, postage stamps or a documentary now underway with Atlantic Productions and Sir David Attenborough. Peter's art will hopefully, in the case of the documentary, form the basis for some sophisticated animation, which again will provide food for thought across a range of intellectual pursuits as well as entertainment for a wider audience.

Figure 11.12 (Left) Early Bushman structure in southern Namibia. Some of the slabs that these hunters used for shelter contained Ediacaran fossils. Ancient Bushman rock art also includes patterns that are reminiscent of these first metazoans. It is likely that this was the first human contact with these most ancient ancestors of ours. (Above) Bushman art carved in the rocks associated with the fossils reflects the shapes of these ancient fossils (photos by P. Vickers-Rich).

The Artist: Peter

In 2003, I had a 'serious' introduction to the Neoproterozoic, via some fascinating conversations with Mikhail (Misha) Fedonkin, who had come to Melbourne at Pat's invitation. 'Inspiring' might be a better word, because this occasion reignited my earlier student interests in the beginnings and diversification of life – matters so very fundamental to evolutionary understanding but that remain mysterious and enigmatic. The manner in which Misha wove his story of research was made all the more enticing to me, not because of the quiet manner in which he compellingly described the scientific issues or revealed the myriad of newly discovered fossils, but because of cultural and philosophical understandings he held dear.

Plans for the book that Misha Fedonkin and Pat were proposing to write were becoming increasingly complex, and they sought my involvement in producing feature illustrations to augment the thorough and ambitious survey that they were planning. Their discussions with me again reinforced the fact that my 'naivety' was going to be just as important as my artistic experience. The factors that underscored this were not apparent to me at the time. Their trust in my ability to contribute to this publication and, indeed, to research in this field, genuinely had me a little bewildered. My experience with invertebrate zoology in general was poor. This I confessed to them, and so what could I hope to provide? Gradually, they introduced me to the Australian Ediacaran

Figure 11.13 Peter (left), Andrey Ivantsov and Mikhail Fedonkin discussing one of Peter's reconstruction sketches at the Paleontological Institute in Moscow, in 2003 (photo by P. Vickers-Rich).

and the Russian Vendian. It was a little time before I realised that my lack of expertise with critters that lacked backbones was not actually going to be pivotal to the issues that confronted the investigations in this field. Inviting me to the wilds of the White Sea region of Russia set the scene for my 'revelation'.

As with my previous introductions to other palaeontological work, Pat ensured that I was to successively gain experience with Precambrian researchers in the field. This plan was to take me to the Flinders Ranges in Australia with Jim Gehling and Mary Droser, to Newfoundland with Guy Narbonne and to southern Namibia with Charlie Hoffmann, but my initiation was to be with Misha and his team in Russia. This was to be further justified by Pat's determination to organise a stamp issue on the Australian Ediacarans with Australia Post – it was all premeditated!

Fundamental differences of approach had to be applied to the investigation of this fossil material. Insofar as illustrating was concerned, I simply started from first principles. Sitting on a driftwood log by the grey shores of the White Sea, I commenced drawing and watercolour painting everything that was before me – indiscriminately. I was desperate to absorb all I could: I wanted to gain familiarity with the sites and the source material being uncovered by the field teams as I sketched. In doing this, I quickly began to see the problems that the preservation of the soft-bodied organisms presented. The researchers were patient with me as I closely examined the poor specimens as well as the spectacular ones. Despite the beguilingly fresh and detailed appearance of the fossils and the incredible preservation of the ancient seabed surfaces on which the organisms were originally buried, what seemed visually apparent was only a small fraction of the story. In order to understand what was coming down to us from such vast passages of time, the poor, fragmentary and associated

Figure 11.14 Spindle-shaped rangeomorphs, *Fractofusus*, from the Mistaken Point area of Newfoundland.

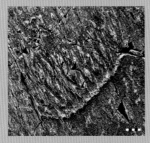

Figure 11.15 Pectinate-shaped rangeomorphs, *Pectinifrons abyssalis* from Mistaken Point area of Newfoundland.

Figure 11.16 Rock outcrop on the Avalon Peninsula of Newfoundland. These sediments seem to have been deposited in deep marine waters, below the level where light could penetrate, and had a significant input of volcanic ash – not a particularly easy environment in which to make a living!

material together with all manner of subtle sedimentological details provided the clues to 'reading' the good specimens. The fossil Ediacarans were highly and variably deformed, soft-bodied impressions in sand or clay. Yes, they were mere impressions, but still remarkable, if only for the fact that they had left any trace of their existence at all.

In one sense, the study process was not all that different from the familiar detective work that palaeontological and archaeological sites demand. But, from the moist, grey-green siltstone cliffs that were literally avalanching around our tents, the rocks were rapidly reverting to grey mud slurries in the rain. In turn, this flushed the sea to a milky grey, blurring the horizon into grey shrouded skies. Everything was a ghost. This was the fundamental difference. Where fossils in the Phanerozoic were somehow real, directly related to tangible, solid forms and structures by virtue of the replacement of their remains with minerals (solid objects turned to stone), the White Sea Ediacarans were apparently *not* being preserved. We were essentially dealing with crumpled masks in subtle, reversed, shallow relief.

Jim Gehling later described the preservation style to me from a sun-baked hill in the Flinders Ranges, as 'the impressions of plastic bags in sand'. Aside from some oxide staining or patches of pyrite deposition, in most cases you could not detect the presence of the fossils on the White Sea stone unless the fossiliferous surface was angled into a position where raking light revealed their subtle shadowed contours. Hard to do on a cloudy day! This was to prove to be the same scenario in Newfoundland, where the fossils were irretrievably attached to expansive, immoveable bedding planes. Beneath the sunnier skies in both Australia and Namibia, prospecting was essentially restricted to early morning or late afternoon. One simply cannot see the specimens in direct light.

Furthermore, the impressions for the greater part of the Russian and Australian sequences were generally only apparent on the sole (the underside) of the bedding surfaces. It seemed everything about the Ediacaran world was upside-down and in negative relief. There was often no counterpart to the impression, because only the surface of the original biomat layer that grew over the sandy sea floor of the time was being cast by the sediment of the burial event. This sediment had apparently lithified quickly, before the biomat had decayed. It preserved the dorsal surfaces of the undecayed or living organisms that lay upon it, together with the top of the biomat. Generally speaking, everything that was not in direct contact with the burying sands on that plane left little or no trace.

Sometimes it was more complex than this, because any more resilient features, within or on either side of the soft organism thus flattened, could imprint a slight impression through to the surface in contact with the entombing sand. The thinly cast relief could, therefore, contain multiple layers of overprinted relief. Internal anatomy might be showing, but in combination with surface details. Any mark or disturbance to the biomat surface layer was cast equally, but in the opposite relief. This, too, could be recorded through the impressions of the biota. Thus, the preserved detail that I was inspecting on every surface was potentially a composite mould of many things; some

CHAPTER 11: *THE RISE OF ANIMALS* 217

Figure 11.17 The Flinders Ranges landscape in South Australia is one of the significant regions where Ediacaran animals have been recovered. It was the fossil material first recovered here by Reg Sprigg in 1946 that led to the realisation that animals had existed prior to the Cambrian (photo by P. Vickers-Rich).

Figure 11.18 *Tribrachidium* from the Flinders Ranges in South Australia, about 2 cm in diameter (photo F. Coffa).

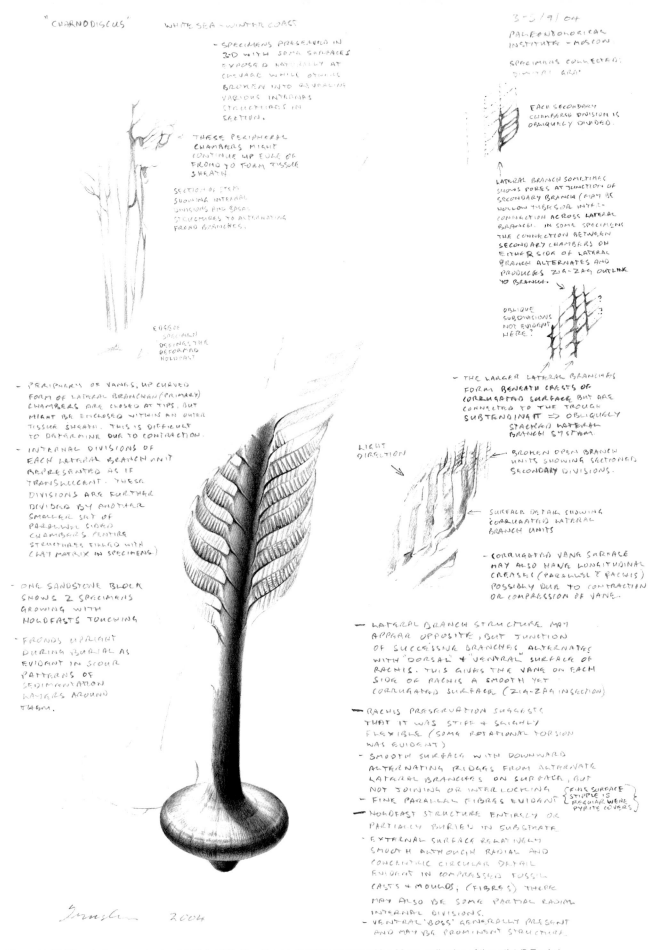

Figure 11.19 '*Charniodiscus* – note sheet' (2004): White Sea, Russia. Graphite on paper, 38 x 28 cm, collection of the artist (P. Trusler). *Charniodiscus* was once thought to be related to the soft corals, but its relationships are now not clear.

in positive relief; most in negative relief. One could not actually 'prepare out' any of the presumed layers because the relief on the underside of the original parted surface was all that remained. In Russia this rock surface was soft and friable, whereas in Australia the quartzite was hard and far more resistant to weathering.

It was a mind game of teasing apart the superimposed and deformed layers of patterns, but this was only where analysis began – it was something that had to be returned to, and reassessed, time and again in breaking the code of the Ediacaran world.

The patterns being revealed were also telling a unique story in so many cases. There were remarkable similarities to modern organisms in some instances, and totally novel body plans in others – symmetries that do not seem to continue beyond the Cambrian boundary and certainly ones that do not exist in biology today.

Two brief examples will demonstrate these opposing trends. For many years the frondose organisms, such as *Charniodiscus* and *Charnia*, have been immediately compared to modern soft corals, to the benthic-dwelling sea-pens in particular. The stems and cylindrical or disc-shaped holdfasts are often understood to be flexible and deformable structures, akin to those of pennatulaceans. The large, blade-like fronds, which were demonstrably held erect in the water column, were similarly divided, or branched, in a manner reminiscent of this group of cnidarians. So compelling is the visual similarity that detailed functional units of cnidarian anatomy have been ascribed to the form of these fossils as well as their biological and ecological homologues. While this may not necessarily be incorrect in any regard, being able to substantiate much of this is proving remarkably difficult. It is not an unreasonable proposition, but it is an untestable hypothesis. How could we test or verify similar ideas?

A second example is the remarkable medallion-like fossil *Tribrachidium*. When I joined the Russian team on the White Sea coast, they were recovering numbers of these beautiful specimens – exquisite for their finely detailed, tri-radial symmetry. To me, their remarkable body form seemed to materialise in the fossil record with no apparent precursors, and disappear just as quickly, leaving no apparent descendants. These were just totally novel and left me in a state of baffled wonder.

We are so reliant on an ordered process of comparison to the known biological framework, as we understand it, that attempting to place many of the Ediacaran fossil forms in any sort of systematic sense has been a brain-teasing exercise. Some taxa have been individually described as belonging to each one of the known biological kingdoms, but a consensus has not been reached. Some researchers have erected new kingdoms in which to place them. In short, we are almost clueless, such is the nature of the preservation and/or the weirdness of the original biota. In the time that I have broadened my study of fossils from the major Precambrian sites, I have come to realise the depth of the problem arising from the perception and recognition of pattern. There is a powerful psychology involved with the process of pattern recognition, and from our very human visual ability to assemble familiar imagery from jumbled patterns. Just as children might see animal shapes in the clouds, or some sci-fi enthusiasts can identify faces from mountains on Mars, scientists, too, can readily identify complex, familiar structures from seemingly random fields of information. For an artist, this capacity is highly developed, and for the scientist it is an invaluable tool of trade. It became clear to me that this strong mental capacity was also likely to be our Achilles heel when dealing with such novel and subtle patterns of moulded relief. Experience quickly taught me that it was so easy to see what I wanted to see, and so difficult to establish a rigorous process to test the validity of this. Where I felt that quite detailed identifications of morphological structures, or taxa, had been dictated more by expectations of what should be present than by tested assessment, I became increasingly suspicious. But how could I improve my own perceptions and counter my tendencies, no matter how naïve, to do exactly the same? This presented me with a unique challenge.

One of the fundamental issues with this moulded preservation style is that it is totally dependent on the grain size of the entombing sediment. Jim Gehling wryly expressed it in terms that I could understand by saying: 'It is like trying to identify someone's portrait from a cast of their face that has been made with ping-pong balls for casting compound. How could you tell if the face had eyes and nostrils?' This is the issue. Even the finest of preservations seldom provide the resolution to identify key biological features.

As the specimens were being gathered and the discussions taking place between all those involved from such a variety of disciplines, I was schooled in ways to eliminate many of the variables that were clouding the analysis. I was on a steep, taphonomic learning curve – that is, learning to understand all of the physical factors that influence the processes of decay and preservation. The nature of the soft-bodied biota, the environment in which they lived, and the circumstances under which they died had come down to us through an enormous passage of time. I had never dealt with fossil material that was so old. Everything that seemed so visually apparent really needed to be reassessed once the taphonomic 'filters' were decoded.

The final revelation came to me a little later, after I had begun comparing the Russian material with its remarkably similar and contemporaneous fauna in Australia. I had the opportunity of prospecting in the Flinders Ranges with the South Australian Museum crew and the Waterhouse Club, a group of serious natural history enthusiasts associated with the Museum. Jim Gehling was giving this group a detailed geological tour of the region. Again, so many members of this expedition brought with them a diversity of expertise and lively discussion. This group was actively assisting Jim and Mary Droser in detailed excavation and study of the ecology that was being revealed with the exposure of expansive sections of the Ediacaran seafloor. Prospecting the low eroded hills had been a process of flipping over the rippled, plate-like slabs of rock to find the fossils imprinted on the undersides. Occasionally, small sections could be pieced back together. The crew had become proficient at jigsaw puzzles, but in order to learn more about the entire palaeo-ecological story, more systematic efforts were needed. Under Jim's guidance, successive squadrons of volunteers had excavated into the hills at key positions. Having mapped each stratum, they removed the successive layers, turned each stone fragment and reassembled them according to their 'jigsaw

Figure 11.20 '*Ediacara*' (2004): field study from Flinders Ranges expedition, June 2004. Watercolour and gouache on paper, 37 x 27 cm, private collection (P. Trusler). An Australian landscape where exposed sandstones that have weathered out from the low hillsides have been tagged by volunteers from the South Australian Museum. Each ribbon marks the location of a fossilised imprint of a soft-bodied organism. The next team of volunteers will record and number each specimen.

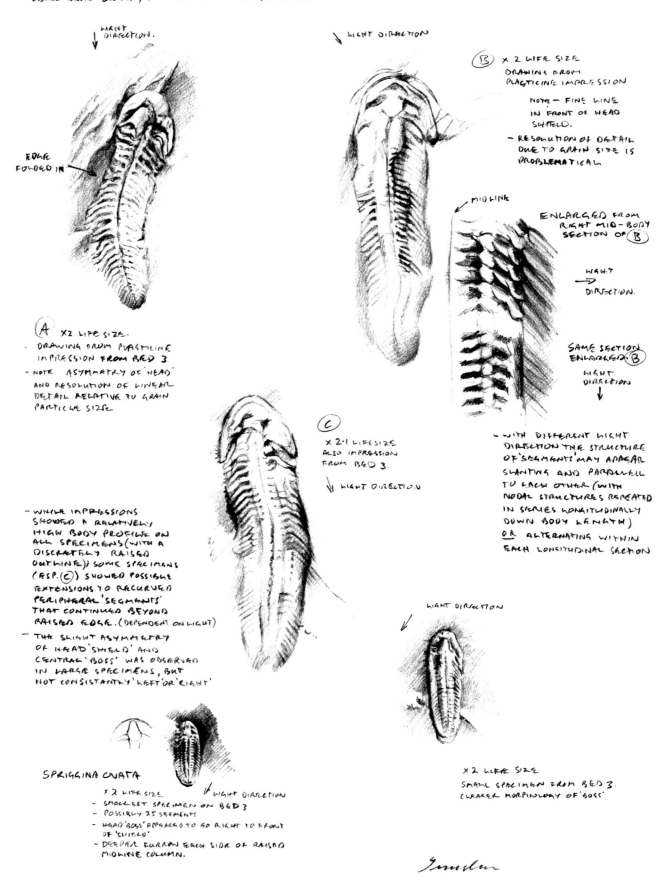

Figure 11.21 'Spriggina' (2004): Flinders Ranges, South Australia. Graphite on paper, 38 x 28 cm, collection of the artist (P. Trusler). *Spriggina* studies recording several different individuals from one area of preserved seafloor. *Spriggina* can reach up to 2–4 cm.

blueprint'. Vertical layers of time could now be spread in horizontal sequences. And so, after 550 million years, these inverted pavements revealed the patterns of colonisation and distribution of the biological communities exactly as they had been prior to each burial event. Storms were the most likely cause of such events – churning up the coasts and estuaries and casting flurries of sand over the shallow sea floor.

Figure 11.22 'Mistaken Point' (2005): landscape of the Precambrian fossil locality, Avalon Peninsula, Newfoundland, NAPC field trip June 2005. Watercolour and gouache on paper, 37 x 27 cm, private collection (P. Trusler).

My first attempt to publicly present reconstructed examples of the Australian Ediacara biota was via the large-format special philatelic issue of 2005, and it was essentially informed by my Australian and Russian experiences. It was due to the unreserved access to the collections and research of the teams in these localities that this was possible. Scientists, students and technicians alike took me into their confidence and shared so much of their unpublished work. They also had the good grace to allow me the opportunity to ask question after question, and indeed to interrogate their work with the observations that I was steadily compiling. For this artist, it has become a great privilege to be able to contribute intellectually. Pat and Tom were certainly well used to my curiosity, and we were comfortable with the process – I entered this new scientific field, however, with less confidence.

There was considerable comfort in my working relationship with Australia Post. The professionalism of their design and production studio had always been inspiring, and I have enjoyed working through projects with such lively creative input from all sides. *Australia's Dinosaur Era* or *Birds of Prey*, for example, might now seem to possess obvious commercial merit, but to my mind sessile organisms that grew and lived on bacterial and algal ooze were in the 'not-for-sale' category. Sure, some of the organisms exhibited beautiful structural patterns, and there was a curiosity factor of sorts. But this was probably a time even before eyes had evolved! Some of the creatures could apparently move or flex, and yet we did not know how. The science was new, complex and much of it hotly debated. This was not without some irony, for the evolutionary setting was one of the dawn of metazoan life, all of which was probably very, very simple.

Worse still, the illustrative issues were also complex. How could I hope to engage a wide audience, where essentially the subject had no recognisable identity? By

Figure 11.23 '*Bradgatia* – partial reconstruction, Newfoundland' (2006). Graphite on paper, 27 x 17 cm, collection of the artist (P. Trusler). The self-similar branching architecture of the rangeomorph is clearly seen in this partial reconstruction of *Bradgatia* from Mistaken Point, Newfoundland. This form also occurs in England. The width of the specimen is about 10 cm.

comparison, worms, sea-slugs or jellyfish seemed to enjoy substantial street-credibility and had personalities to go places! The Ediacaran organisms seemed impossible! I underestimated everyone: Pat's determination and communication skills; Australia Post's ability to grasp the international scientific significance and long-term cultural value of such palaeontological sites; the marketing team's ability to divine the potential interest of a novel subject to a young public and take the risk. Above all, I undervalued everyone's genuine inquisitiveness, especially that of children.

Happily, the reception of this project spurred me on.

Colleagues put the word out: 'Trusler has to see the Newfoundland biota.' Pat immediately provided for my fares. Guy Narbonne quickly found me a place on a field trip that he and Jim were conducting. In a matter of days I was standing atop a precipice, gazing out towards the fog banks rolling in from the western Atlantic. The *Titanic* went down out there. The lighthouse that received the distress communications during this famously fatal episode of 20th century history was standing on an identical precipice a few kilometres away. All along this stunning peninsula, the jagged grey cliffs were composed of regular tilted lines of alternating mudstone and ash, representing a remarkable sequential record stretching back to the earliest days of the Ediacaran Period. Distinctively, the sediments had been deposited on an ancient ocean floor at great depth. Life had flourished and died there in the same black world that was now the tomb of that ill-fated ship and her passengers. The slow warping and heaving of the Earth's crust had since raised these ancient layers from the depths, but with each pounding wave the sea was slowly reclaiming them.

On this ancient seafloor early metazoan life had been smothered by massive volcanic ash plumes that drifted seawards and settled to the depths. Today, as this consolidated ash is eroded from the top of each layer, the undersides of the organisms that were pressed into the mud can be seen as shallow impressions all over the surfaces. However, such 'snapshots' of these ancient, distinctive communities had been substantially distorted. Tectonic compression of the sediments had skewed the original proportions of these rocky surfaces such that an original circular shape was now elliptic. A cast of these impressions provides the original positive form of that part of the organism that was in contact with the seabed, but its shape needs to be retro-deformed to get the real picture. Once the degree and direction of alteration is assessed, researchers can correct this graphically, photographically – or, better still, digitally.

In a similar manner to the Russian and Australian sites, the strength and direction of the felling current could be determined from the constancy of direction in which the upright growing, frond-like organisms had been flattened. This could also distinguish those species that were randomly orientated – perhaps fully sessile forms or those with free-swimming or floating lifestyles that had ultimately come to rest on the ocean floor. Ripple marks and subtle surface contours about each specimen could be used to demonstrate the strength and direction of the prevailing currents too. The varying degrees of preserved detail of the specimens themselves gave clues to the deformation of their original living form. All this had to be factored into my work. The graphic process became a tool for analysis in its own right.

Figure 11.24 '*Beothukus* – Newfoundland' (2008). Gouache on paper, 27 x 17 cm, collection of the artist (P. Trusler). Reconstruction of another rangeomorph form described from Spaniards Bay, Newfoundland. The length ranges from 3–9 cm.

The discussions among the diversity of experts on this field trip provided me with ample things to contemplate, but Guy and his student, Mark LaFlamme, had yet another card to play.

They introduced the group to a site at Spaniards Bay, where a different sedimentary process had preserved mostly small specimens in true three dimensions. Down-slope slurries of mud had buried the organisms where they grew, and minerals had quickly infused their bodies before they had significantly deteriorated. They were then further hardened and protected by the concretion of minerals that continued to grow around them. At last, real fossilisation! Well, not quite – because the erosion mechanism that had revealed the frond fossils on a section of exposed bedding plane was relentless wave action pounding stones over the surface. This had largely hollowed out the delicately-branched, petrified specimens from within the concretions. The degree to which this had happened was sufficiently varied to provide layered views from one side of a specimen to the other. The structures were, therefore, variously positive or negative, revealing undistorted surfaces as well as internal anatomy. Again, casting them in latex was the key to investigating large numbers of the specimens. The advantage here was

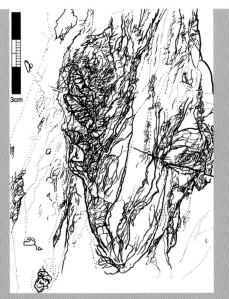

Figure 11.25 Digital line graphic recording the complex moulded surface topography of a rangeomorph specimen from Spaniards Bay, Newfoundland (2006). All significant features of the surface have been recorded irrespective of whether they might be of fossil or some other origin (P. Trusler).

Figure 11.26 Same specimen as in Fig. 11.25 (2006), showing the analysed surfaces belonging to the fossil as colour-coded for differing types of surface relief. This technique made it easier to investigate how the different components of the specimen were being preserved, which in turn allowed for a reconstruction of the true structure and dimensions of the living organism (P. Trusler).

Figure 11.27 '*Pteridinium* – Namibia' (2006): Precambrian fossil locality; inset: weathered quartzite, Upper Kliphoek member of the Dabis Formation of the Nama Group, with *Pteridinium* specimens – 'living stones', *Lithops* species – and lark-like buntings, *Emeriza impetuani.* Watercolour and gouache on paper, 37.5 x 27.5 cm, private collection (P. Trusler). The segmented, twisted, fusiform shapes of *Pteridinium* fossils are visible on the boulder surfaces strewn over the southern Namibian desert. The flat-topped escarpments on the horizon are capped by limestones marking the environmental and evolutionary transition to the Cambrian seas. The modern drought-adapted flora and fauna flourish here in good seasons. Birds are the size of a house sparrow, for scale.

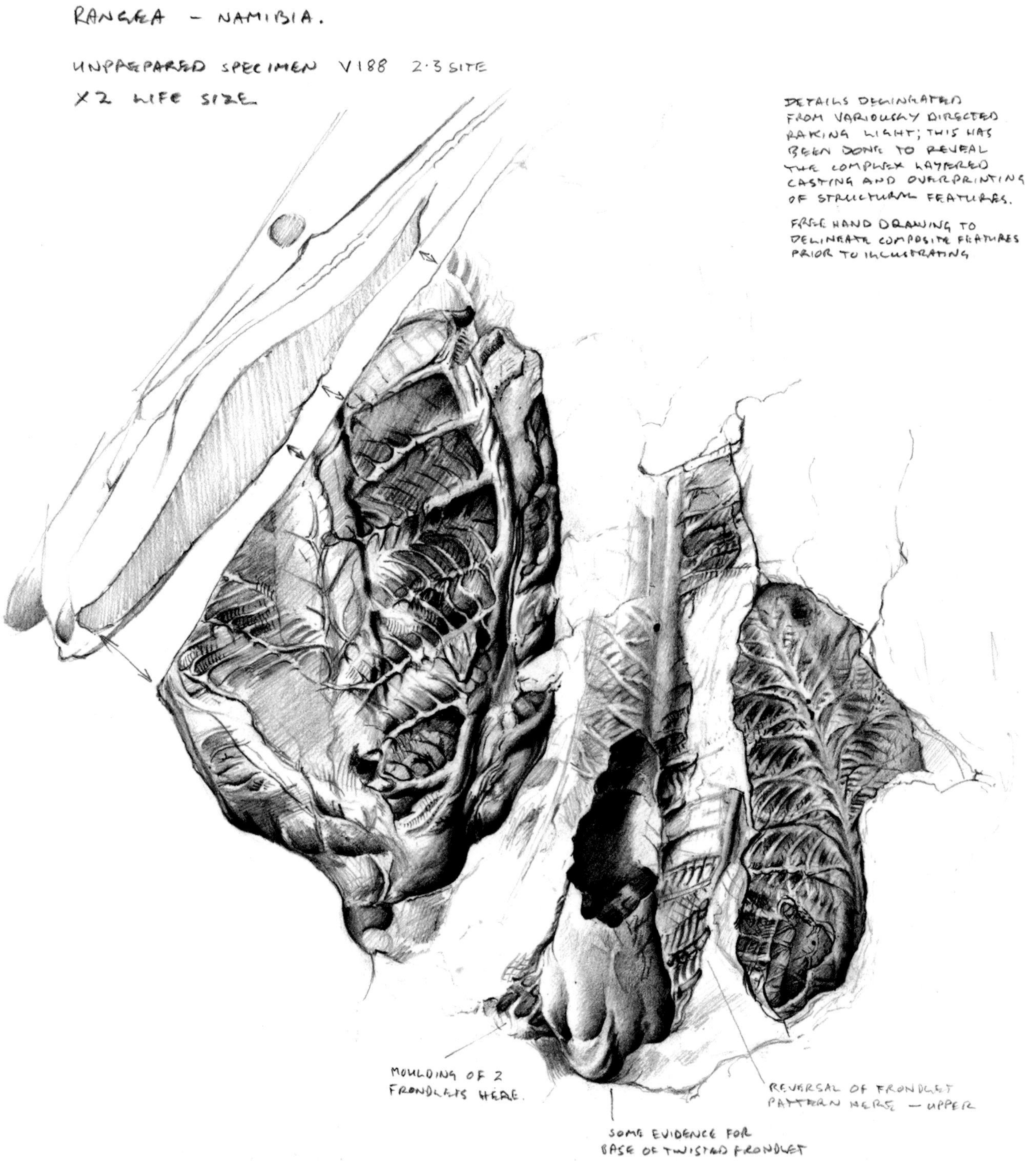

Figure 11.28 '*Rangea* – Namibia' (2008). Graphite on paper, 38 x 28 cm, collection of the artist (P. Trusler). Sketch notes taken of *Rangea* specimens from Namibia.

Figure 11.29 Precambrian trace fossils from the Flinders Ranges in South Australia. These only occur on the surface of seafloor sediments and within the biomat. Animals did not burrow more deeply until the Cambrian. This may have resulted from the development of predators with eyes. Defence strategies for evolving animals that were preyed on included digging deeply for escape and developing a hard shell for protection – these attributes quickly became widespread in times younger than 542 million years ago. (photo by S. Morton). The width of the specimen is 11 cm.

that because the preservation relief was not originally cast by sediment, the detail was not limited by sediment grain size. Despite the small size of the specimens, their significantly superior preservation provided a guide to interpreting the majority of the Avalon Peninsula fauna.

I could now colour code the different surfaces in a diagrammatic technique to re-establish the three-dimensional composition of the entire anatomy. We were steadily progressing!

In the interim, Pat and her Russian colleagues, Mikhail Fedonkin and Andrey Ivantsov, were being assisted by the National Geographic Society and the Namibian Geological Survey to search for new material in Namibia. Unique fossils had been known there since the early 20th century and several sites across the extensive exposure of Precambrian rocks continued to provide major contributions to the science. Dolf Seilacher and his colleagues, in particular, had stirred the debate and broadened our understanding – in part through their spectacular discoveries in Namibia. Importantly, the geology spanned the latter Ediacaran and into the Cambrian. Better still, it contained massive quartzites that had preserved and moulded specimens three-dimensionally. These were formed by huge slumps of sand that had cascaded down marine valleys and gutters – avalanches that took some of the sea-floor life with them as they went and covered others. Examining these would give me fresh insights, and provide comparisons to the new locations in Australia and Russia that showed similar geological settings and preservations. In Namibia the rock was hellishly hard, and those specimens that were not weathered naturally were nearly impossible to further expose. I needed to cope with substantial distortions by virtue of their tumultuous burial. Not only were the specimens frequently intertwined or broken apart but much of the documented material was also collected as rubble. Consequently, it was seldom possible to glimpse more than a portion of any one organism, and it involved too many assumptions to place them in their original life setting. We still needed 3-D material *in situ* and of a different preservation that could be more reliably prepared out of the sediment. Pat found the specimen that gave us the lead, so back we went to successfully continue our research! It was high time the artist did some real digging. And dig I did!

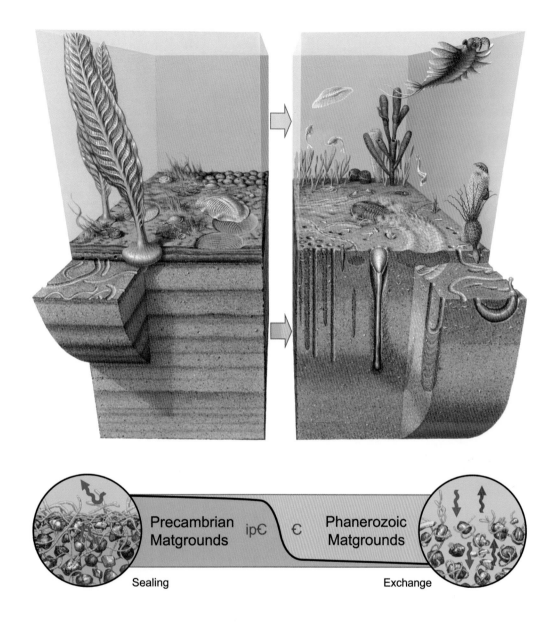

Figure 11.30 'Agronomic revolution' (graphic illustration, *The Rise of Animals*, Fedonkin et al, 2008). Gouache on paper, 28 x 38 cm; collection of the artist (P. Trusler, based on a concept developed by Dolf Seilacher). The differences in the make up of the biota and the sediments on the sea floor in Precambrian times (left) and Phanerozoic times (right), the latter being younger than 542 million years ago or more recent. Animals began to burrow and form hard shells in Cambrian times, making the seas very different places than they were in older times.

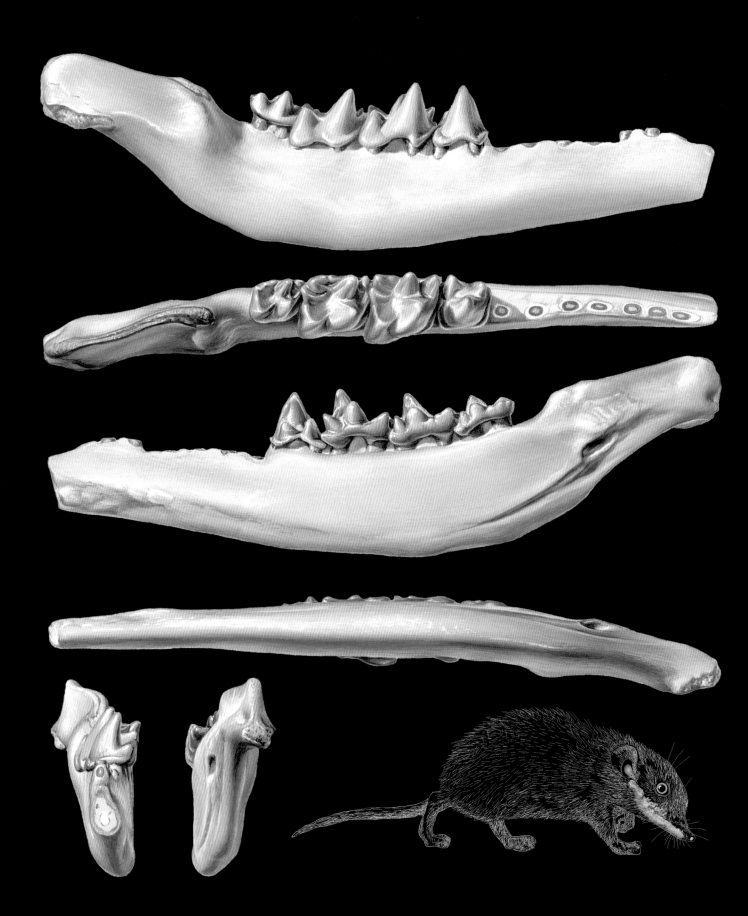

12

Chinese and Australian Mesozoic Mammals

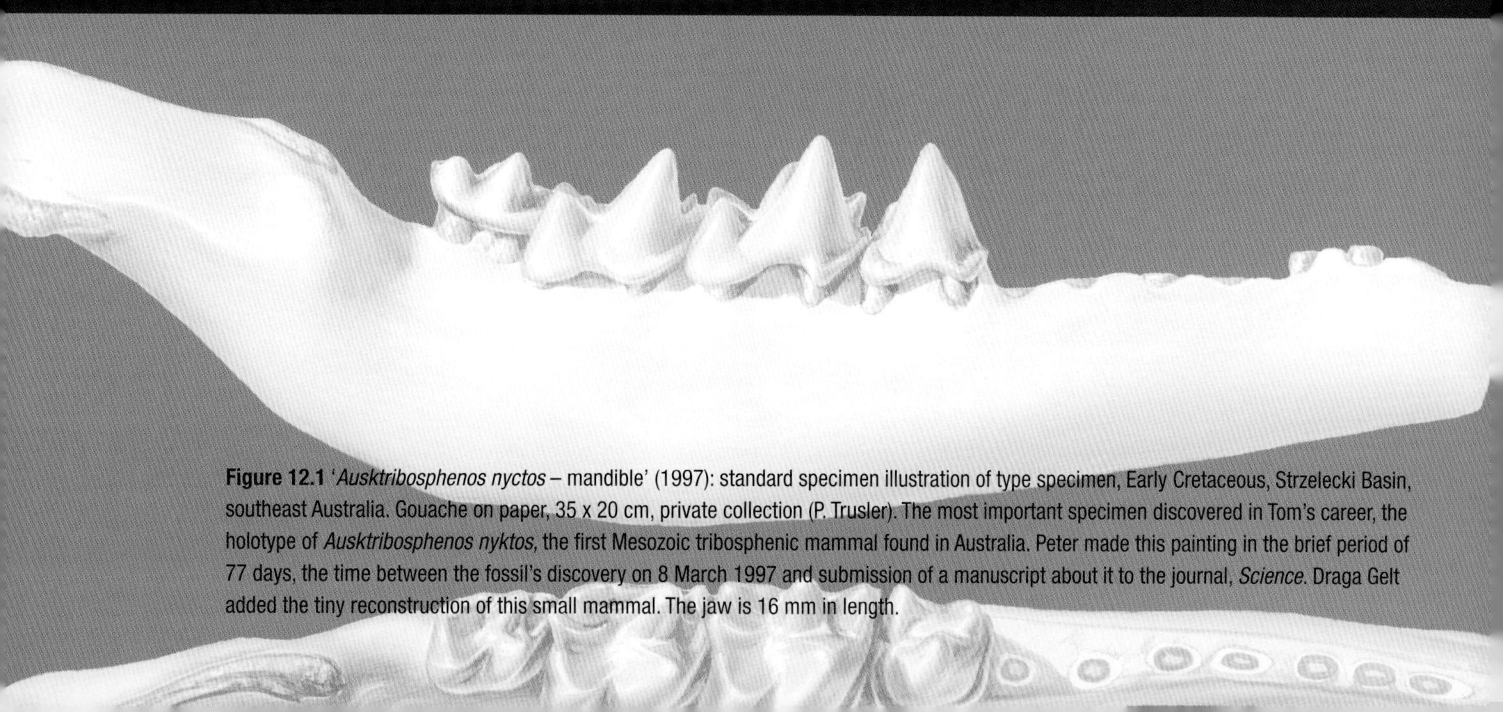

Figure 12.1 '*Ausktribosphenos nyctos* – mandible' (1997): standard specimen illustration of type specimen, Early Cretaceous, Strzelecki Basin, southeast Australia. Gouache on paper, 35 x 20 cm, private collection (P. Trusler). The most important specimen discovered in Tom's career, the holotype of *Ausktribosphenos nyktos*, the first Mesozoic tribosphenic mammal found in Australia. Peter made this painting in the brief period of 77 days, the time between the fossil's discovery on 8 March 1997 and submission of a manuscript about it to the journal, *Science*. Draga Gelt added the tiny reconstruction of this small mammal. The jaw is 16 mm in length.

In 1982, Prof. Minchen Zhou (or Chow), then director of the Institute of Vertebrate Paleontology & Paleoanthropology, spent three months living with Pat and me in Australia. Shortly after he arrived, he asked me to describe a tiny mammal lower jaw from Sichuan, China, which he had brought with him to study. I was really chuffed to be asked, because I had never previously had the chance to study and publish on such a fossil, although both Pat and I had put in much time in the field collecting them for other researchers in North America.

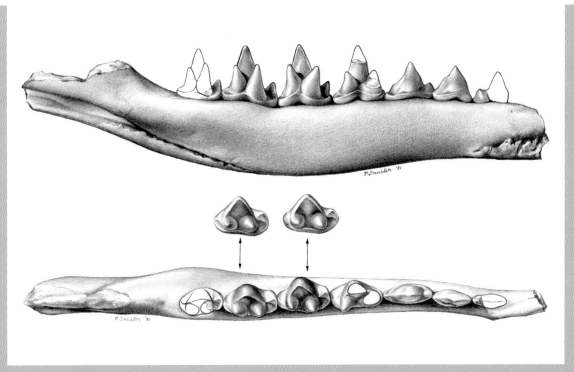

Figure 12.2 '*Shuotherium dongi* mandible' (1981): standard specimen illustration. Graphite on illustration board, 2 panels 10 x 20 cm, private collection (P. Trusler). The first known specimen of a major group of mammals (characterised by pseudotribosphenic teeth), the yinotheres, as first illustrated in Chow and Rich (1982). This small mammal, *Shuotherium dongi*, from the Middle or Late Jurassic of China, was unique in having 'backwards' molars. The jaw is 11.5 mm long.

When I first put the specimen under a microscope, I was completely baffled. The various structures on the tooth did not fall neatly into a familiar pattern. What was this animal, and what in the world did it do with those peculiar teeth?

After pondering awhile, I suddenly realised that with a 180° flip of the morphology, the pattern was familiar to me. In essence, it was structurally back to front from the pattern that almost all living mammal molars exhibit, or that shown in the teeth of the ancestors they are descended from. In this common pattern, on lower molars there is a pillar in front of the tooth against which the upper molars shear or slice the food. In the rear of those lower molars is a basin where food is smashed by a cusp on the corresponding upper molar. In the Chinese jaw, the lower molar teeth had the crushing basin in front and the slicing pillar behind! This was certainly an alternative way of accomplishing the two jobs of slicing and crushing food on a single tooth – a dual dental function that no reptiles ever evolved. It was intriguing to me at the time that the functional path taken by this ancient Chinese mammal was unsuccessful in the long run. Why was it not equally as successful as the alternative structure adopted by the majority of living mammals and their ancestors? But then, there was no indication of what the structure of the rest of these animals was like. Long-term evolutionary success or failure may well have been owing to other differences between the two groups of which we had, and still have, no knowledge.

TELL-TALE TEETH

The key to understanding fossil mammals generally, and certainly those from the Early Cretaceous of Victoria, Australia, has always been the structure of their teeth. The teeth, because of their durability, are often the only part of a mammal preserved in a fossil specimen. (However, a bit of palaeontological folk wisdom goes: 'What is the best fossil to have?' 'The one you've got.') Mammalian teeth are complex structures that vary markedly from group to group in their cusp patterns and shapes. So, teeth are favoured objects of study by both palaeontologists and students of modern mammals, neontologists.

When tribosphenic molars meet as a mammal chews, the talonid (or rear part of the lower tooth) acts as a basin that receives a cusp on the upper molar. This action crushes the food. At the same time, the anterior and posterior edges of the trigonid slide pass the near vertical surfaces on the upper molars, slicing food like the blades of a pair of scissors. Except for the monotremes – the egg-laying platypuses and echidnas – all living mammals either have this pattern or are descended from ancestors that did. The pseudotribosphenic teeth of *Shuotherium* operated in much the same way except that the placement of cusps and basins on the lower molars, where the crushing and slicing took place, were reversed, i.e. back to front. In evolutionary terms, the tribosphenic pattern has been highly successful in that it was a pattern common to almost all living mammals at some stage in

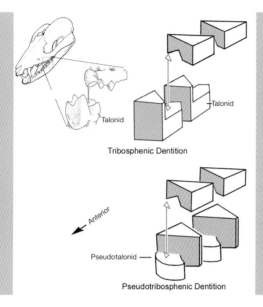

Figure 12.3 Comparison of the tribosphenic and pseudotribosphenic molar tooth types. The two molar types both crush and slice the food. However, on the tribosphenic lower molars the crushing is done by the talonid at the rear of the tooth, while on the pseudotribosphenic forms the crushing is done by the pseudotalonid at the front of the tooth. The taller triangular pillars, the trigonids, on both tooth types carry out the slicing function.

their ancestry. In contrast, for some unknown reason, the pseudotribosphenic pattern was an evolutionary dead end. This was just one of many 'evolutionary experiments' that arose among mammals during the Mesozoic Era, which allowed many early mammalian groups to flourish for a time and then disappear forever.

Once the manuscript on this specimen was completed, the next step was to send it to professional colleagues for comment, a customary process before formally submitting a manuscript to a scientific journal. I distributed it together with photographs of the specimen, to a number of fellow researchers. One was Walter Kühne, who was as intrigued by this remarkable specimen as I was. He commented that if I wanted the specimen to be mentioned in textbooks and other secondary literature, it was critical to have an artist's illustration of it to accompany the photographs. So, Peter was immediately involved.

Peter's illustrations consisted of not only a pencil sketch of the Chinese specimen, much as portrayed in the photographs, but also diagrams of the teeth, with various structures clearly identified. One diagram was prepared in colour – the only time I have ever published a technical illustration in that manner. Colour was applied, following the style of a seminal work by Arthur ('Fuzz') W. Crompton, a respected researcher in the field of mammalian dental function, in order to maximise the ease of interpretation of the novel structures present in the dentition of the Chinese specimen. In order to facilitate this art, Peter, Minchen and I used the only microscope available at the time, which we pushed to the limits of its capability! As a result of the barely passable optics,

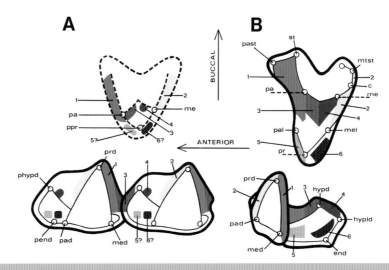

Figure 12.4 Diagrammatic comparison of the molar dentition of *Shuotherium dongi* (A) with that of a tribosphenic mammal (B), as published by Chow and Rich (1982). The upper molars are in the top row and the lower molars in the bottom row. This is a 'food's-eye view' of these teeth. The tooth of *Shuotherium dongi* is backwards, in the sense that the pseudotalonid is in front of the trigonid, while in the tribosphenic mammal the structure corresponding to the pseudotalonid, the talonid, is behind the trigonid. The corresponding colours on the upper and lower teeth show where surfaces on those teeth make contact with one another. The upper molar in *S. dongi* is a hypothetical reconstruction, because the tooth was unknown when Peter drew this diagram.

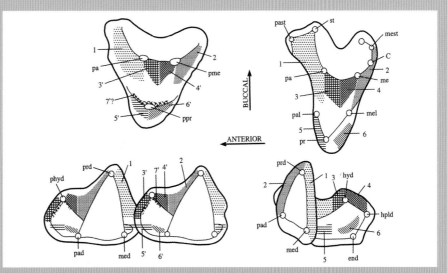

Figure 12.5 Diagram of the upper and lower teeth of *Shuotherium* as published by Wang et al. (1998). The principal difference between this diagram and our earlier one, besides being in black and white rather than colour, is that the upper molar of *Shuotherium* is based on an actual fossil rather than being hypothetical. The upper molar of *Shuotherium shilongi*, upper left, is 2.4 mm wide at broadest; each lower molar is about 1 mm in length.

in some cases we had to infer the presence of some critical structures, tentatively identifying them, not wanting to over-interpret the data. These tentative inferences were made clear with question marks on the diagram. The structures present on the lower molars we had allowed us to provide a hypothesis about the structure of the upper molars of this new mammal, of a type unknown at the time.

Sixteen years later, when Yuanqing Wang and his colleagues had occasion to re-examine our fossil with a decidedly better binocular microscope, they confirmed that our hypothesised structures were indeed present. Their paper described an actual upper molar of the same genus we had named *Shuotherium*, from the same fossil locality as the specimen that we had reported on in 1982. This upper molar was referred to a different species, *Shuotherium shilongi*, rather than to the same species as the lower jaw, *Shuotherium dongi*, because the upper molar was seemingly from a larger animal than would be expected from the size of the lower molars of our *S. dongi*.

By pure coincidence (as far as I know), the same figure number (Fig. 6) was chosen for a dental diagram in Wang et al. (1998) that was all but identical to one used in Minchen's and my paper in 1982. In our paper, Fig. 6A was of a hypothetical upper molar of *S. dongi*. In the 1998 paper, Fig. 6A was of an actual upper molar of *S. shilongi*. Gratifyingly, the differences between the prediction and the actual were remarkably trivial!

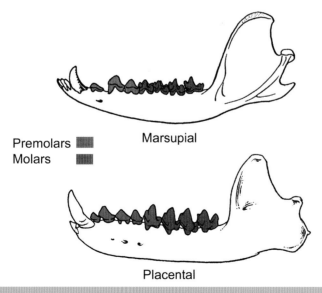

Figure 12.6 Comparison of the molars of marsupials and placentals. Typically, placentals have three molars and marsupials have four. The most posterior lower premolar of a marsupial is much like the more anterior lower premolars and quite unlike the lower molar immediately behind it. In placentals, the most posterior lower premolar is intermediate in form between the lower premolars in front of it and the lower molars behind.

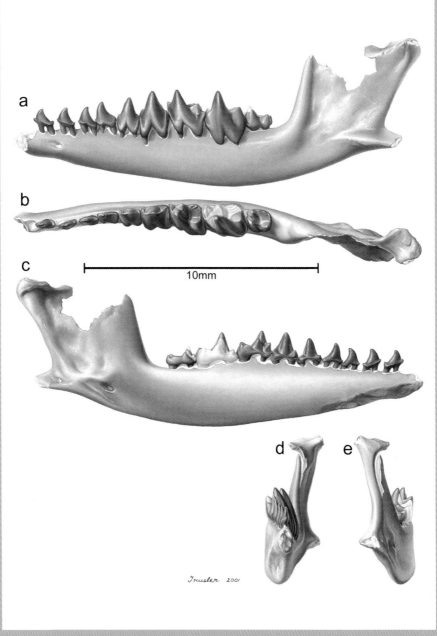

Figure 12.7 '*Bishops whitmorei* – mandible' (2001): standard specimen illustration of type specimen, medial and occlusal views; Early Cretaceous, Strzelecki Basin, southeast Australia. Gouache on paper, 35 x 25 cm, private collection (P. Trusler). Some regard *Bishops* as a placental mammal, which would be a surprisingly early occurrence for this group in Australia; more material is needed for study to confirm its relationship. One thing is for sure: it is neither a marsupial nor a relative of the platypus, i.e. a monotreme.

When Pat and I decided to immigrate to Australia in 1976, Pat was motivated by the desire to continue to study the fossil birds there. I wanted to investigate the first 80 percent of the history of mammals on this continent, which at the time was totally unknown. After 21 years of trying, with the aid of numerous volunteers and Pat's support as a co-investigator, I finally began to make progress toward achieving that goal. On the morning of 8 March 1997 a volunteer digger at Inverloch, Nicola Sanderson (née Barton), discovered a tiny mammal jaw. This time, instead of the fossil coming from far away in China, it came from sediments less than 150 kilometres from our home. This, for me, was the culmination of an effort, begun in 1978, to systematically search coastal outcrops of Early Cretaceous age (105–120 million years ago) along the south coast of Victoria. And I knew that the hardest fossil to find is the first one. This step had now been accomplished. Now, aware of the physical conditions in which this treasured fossil was preserved, I implemented an appropriate systematic search strategy to find more. Over the subsequent dozen years almost 50 more lower jaws were discovered. These finds led to a number of collaborations with Peter to illustrate these rare fossils, both for research purposes and for public presentation.

The fossil found by Nicola was as exciting as the Chinese jaw I had described 15 years before. We interpreted it as being a placental mammal. Others subsequently proposed that it was the first known representative of a mammalian group, confined to the Southern Hemisphere, which vaguely resembles placentals in the structure of its lower jaw and teeth but is not one.

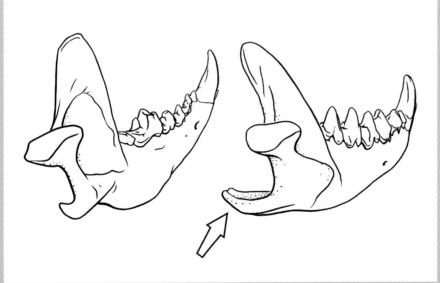

Figure 12.8 A comparative rear view of the jaws of a placental mammal and a marsupial (P. Trusler). Note the presence of an inflected angle on the marsupial (indicated by an arrow) and the lack of one on the placental.

Figure 12.9 *Bishops whitmorei*, an unusual fossil mammal from the Early Cretaceous of southern Victoria. Figs. A–C show the fossil in the rock, during preparation and finally fully prepared and mounted on a pin. Fig. C shows it prepared and just how tiny it is (photos by S. Morton).

A name has even been given by three colleagues (Zhenxi Luo, Richard Cifelli and Zofia Kielan-Jaworowska) to this group that assumes they are non-placentals but convergent on them: Australosphenida. This name reflects their exclusively Southern Hemisphere occurrence.

The debate continues. What makes these creatures appear to be either placentals or mammals convergent on placentals are the fact that they have tribosphenic molars, plus three features that set them apart from the other group of mammals with tribosphenic teeth, the marsupials. Those three features are: (1) they have only three molars, not four; (2) the most posterior premolar is intermediate in form and complexity between the more anterior premolars on the one hand and the molars on the other (that is, it is submolariform) rather than being much more similar in shape to the premolars in front of it than the molars behind; and (3) they lack a structure called the inflected angle on the lower jaw.

Either way, the discovery by Nicola of *Ausktribosphenos nyktos* (which translates as 'the southern Cretaceous mammal of the night with a tribosphenic condition'), together with the closely related *Bishops whitmori* found only a few metres away two years later, revealed a previously unknown chapter in mammalian evolution. Following Walter Kühne's excellent advice 15 years earlier, I wanted to include a painting by Peter of Nicola's important specimen in the resulting publication. However, I hesitated to make this request of him because he was already busy with other commitments. When, despite those misgivings, I asked Peter if he would make the illustration, the memorable part of his response was, 'I would be honoured'. And he soon did it.

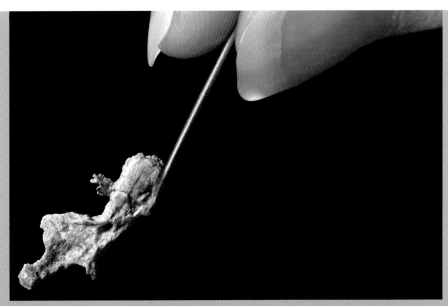

Figure 12.10 The lower jaw of an ancient monotreme, *Teinolophos trusleri*, from the Early Cretaceous sediments at Flat Rocks to the southeast of Melbourne (photo by S. Morton).

Seventy-seven days after the fossil was found by Nicola, a manuscript was sent off to the journal *Science* with Peter's completed illustration. We had hoped to publish this art with stereo photographs of the jaw, the latter beautifully produced by Steve Morton. Because of space constraints *Science* would only allow one form of illustration, so Peter's work it was. Steve Morton's photographs, however, have been used extensively in other research papers and in the popular media, often in company with Peter's illustration of the same specimen.

Another of the nearly 50 jaws was a form we named *Teinolophos trusleri*, which translates as 'Trusler's expanded ridge animal'. It is a prime example of how scientific concepts can change over time. When first described, *T. trusleri* seemed related to the group of mammals out of which the marsupials and placentals arose – the eupantotheres. This assignment held until we asked Charles ('Chuck') Schaff to come to Australia, specifically to prepare one of our Cretaceous mammal jaws from Victoria that was particularly intractable and yet delicate. After working on it for a few days, Chuck asked if he could have some R & R (rest and relaxation) by preparing further the one tooth of the then only known specimen of *T. trusleri*. Soon he was finished, and as everyone examined this newly enhanced specimen, we all remarked how similar it was to the monotreme *Steropodon galmani*. *Steropodon* was known from a beautiful, uniquely Australian fossil: an opalised jaw displaying all the colours of the rainbow, from slightly younger rocks at Lightning Ridge in northern New South Wales. When this fossil appeared on the cover of the scientific journal *Nature*, it was illustrated conventionally in reflected light, but also with transmitted light – in essence it was a natural cast formed of hydrated silicon, optically quite similar to glass.

In exposing more of this specimen, Chuck's R & R had brought about a new development. Science had moved on, as it delightfully does. Although there are today around 50 specimens of Cretaceous mammals known from Victoria, the effort to collect them has been significant, spanning 15 field seasons thus far. During that time, our field crew size has been between about half a dozen and 15. Many who come and contribute weeks or months of effort never have the satisfaction of finding even a single mammal jaw. However, some people who have returned year after year to the dig have been lucky enough to find more than one. The person whom we hope will be the first 'ace', the finder of five mammal jaws, is Mary Walters. She has so far found four. The fourth one was certainly the prize of the lot. It was the first specimen of a group of extinct mammals not previously known from Australia, the multituberculates. In honour of her discovering that specimen, the fossil was named *Corriebaatar marywaltersae*. The generic name honours Corrie Williams who, when accompanying Pat and me to Patagonia in 1997, found the second multituberculate known from that continent. And the *baatar* is Mongolian for hero or heroine; it has been appended as a suffix to many multituberculates not only from Asia, but also North America, so extending the usage to Australia is quite appropriate. Peter's beautifully rendered image of Mary's find had a prominent place in our research paper on this specimen.

In the years that followed, Peter's illustrations of all the other Early Cretaceous mammal jaws found on the Victorian coast appeared, soon after they came to light, in four more publications. In this way, illustrations by him now exist for half the Mesozoic

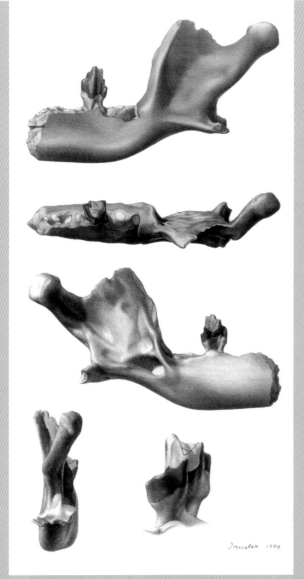 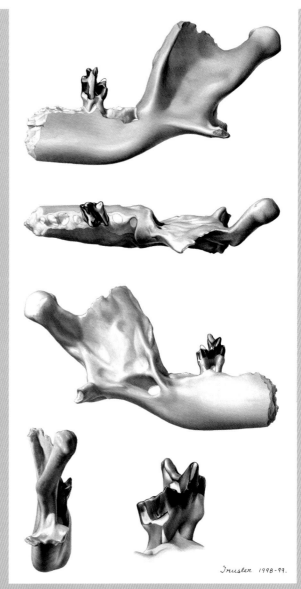

Figure 12.11 '*Teinolophos trusleri* mandible' (1998): standard specimen illustration of type specimen, Early Cretaceous, Strzelecki Group, southeast Australia. Gouache on paper, 35 x 20 cm, private collection (P. Trusler). Changing interpretations of a fossil necessitate improving depictions of it. Originally, the single tooth remaining on the first fragile specimen of the Cretaceous mammal named in honour of Peter, *Teinolophos trusleri*, was illustrated as shown here. At that time, it was tentatively assigned to the eupantotheres, an entirely extinct group of Mesozoic mammals. The length of the jaw is 13 mm.

Figure 12.12 '*Teinolophos trusleri* mandible – revised' (1998–99): standard specimen illustration of type specimen, Early Cretaceous, Strzelecki Group, southeast Australia. Gouache on paper, 35 x 20 cm, private collection (P. Trusler). Later the following year, the risky removal of the remaining rock covering the crown of this tooth was carefully undertaken by Charles Schaff, clearly revealing that the tooth was that of a monotreme. The length of the jaw is 13 mm.

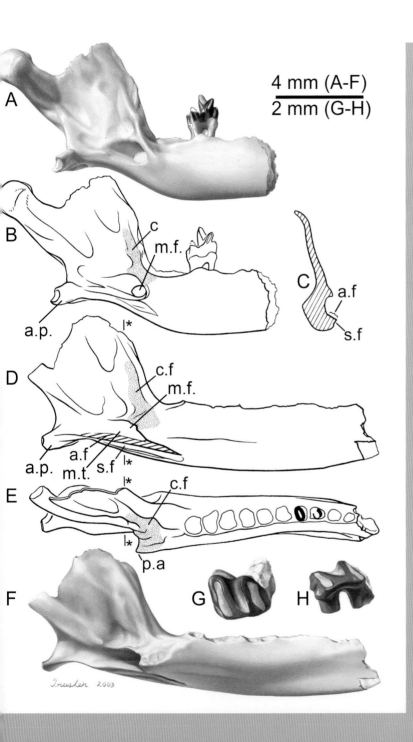

Figure 12.13 *Teinolophos trusleri* continues to be the focus of research into the origins of monotremes. This figure was rendered by Peter to support evidence that there were a number of bones in the lower jaw of the living individuals, based on the presence of grooves in the one bone of the jaw preserved, the dentary. The explanatory caption for the original publication was: '(A) Medial view of holotype of *T. trusleri*, specimen NMV P208231. (B) Diagrammatic medial view of NMV P208231; the stippled area indicates the position of the fused coronoid bone. (C) Cross-section of mandible of referred specimen of *T. trusleri*, NMV P212933; position of cross-section is indicated in (D) and (E) by lines terminated with asterisks. (D) Diagrammatic medial view of NMV P212933. The stippled area indicates the position of the contact facet for the coronoid bone. Diagonal lines indicate the flat facet interpreted as a contact surface for the angular bone. (E) Diagrammatic dorsal view of NMV P212933. Traces of roots of a molar can be seen in alveoli three and four. (F) Medial view of NMV P212933, rotated slightly medially toward the viewer. (G) Occlusal and (H) medial views of isolated lower molar associated with dentary, NMV P212933. Abbreviations: a.f. – angular facet; a.p. – angular process; c. – coronoid; c.f. – coronoid facet; m.f. – mandibular foramen; m.t. – mandibular trough; p.a. – posterointernal angle; s.f. – splenial facet.' The length of the jaw is 13 mm.

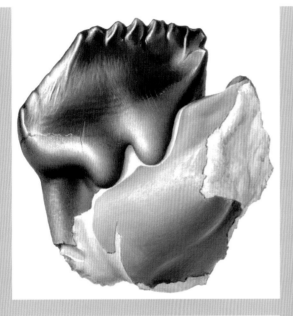
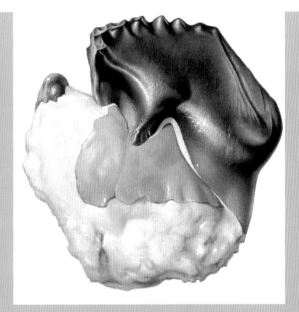
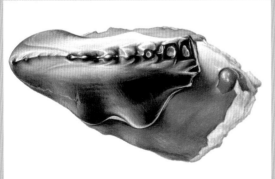

Figure 12.14 'Multituberculate premolar' (2006): standard specimen illustration of *Corriebaatar marywaltersae*, lower, first pre-molar, Early Cretaceous, Strzelecki Group, southeast Australia. Gouache on paper, 35 x 20 cm, private collection (P. Trusler). A single tooth of the only multituberculate mammal known from Australia. Although this primitive group is rare in the Southern Hemisphere, it has also been recorded from South America and Africa. Multituberculates occupied niches that were later taken over by rodents, but rodents did not exist in the Cretaceous and so these now extinct mammals prospered then. Labial view (upper left), lingual view (upper right) and occlusal view (bottom). *Corriebaatar marywaltersae* tooth is about 3.4 mm in length.

mammals presently known from Australia. In working together with me, Peter did far more than provide an illustration. In order to obtain the most detailed and accurate rendering of those jaws, he needed to fully understand their function as well as their systematics, not just as inert objects but also in their full scientific context. Peter queried me constantly about features he considered significant. This not only made me think further about the morphology of the specimen but, from time to time, made me aware of a significant feature that otherwise would have escaped my notice.

One of the chief advantages of Peter's detailed illustrating of specimens is that he can – in a manner that is not possible with an electronic image – correct for breakage and distortion of specimens by tectonic forces acting on the rocks in which they were buried for more than 100 million years. Peter can also emphasise subtle features that are difficult to visualise in two-dimensional or stereophotographs, even those produced with elaborate lighting techniques.

Not many objects that humans handle frequently survive for even a thousand years. So, when these Mesozoic mammals were extracted from the ground, 99.999 percent of their existence as such an entity is probably over! In the case of the Victorian specimens, because their fragility, even that 0.001 percent of their continued existence in the future is unlikely. For this reason, if for no other, every possible way of documenting each of these precious specimens should be made. It is quite likely that, in centuries to come, Peter's illustrations in our joint publications, surviving in libraries around the world, will be the only record of our precious mammals.

The Artist: Peter

Finding Mesozoic mammals in Australia had been Tom's Holy Grail. Following the fourth season of the Strzelecki Group excavations at Flat Rocks, each annual excavation has revealed one or two tiny crushed, isolated mammal jaws. To this day, this seems to be a really paltry harvest after years of hard labour. However, it must be borne in mind that this one site has yielded about 85 percent of the Mesozoic mammal specimens known from Australia, and these have provided unexpected revelations. As yet, there is no known alternative locality where one can go on the Australian continent and confidently expect to collect such fossils.

Tom and I had collaborated closely before on his study with Minchen Zhou (Chow) of the Chinese Mesozoic jaw specimen *Shuotherium.* This experience had prepared me 'artistically' for dealing with these new small-mammal discoveries. Tom was, of course, immediately struck by the morphology of these first fragmentary jaws and their few proudly persistent teeth. They were not what I was expecting either, being overly accustomed to modern marsupial teeth and those of their not-too-distant ancestors. Except for the brief encounter with the Chinese *Shuotherium dongi* 15 years earlier, the scientific 'ball game' on the Mesozoic mammal field was a new one for me!

Such 'naivety' can at times be an advantage. In the appropriate context, fresh eyes, or those from a different personal perspective, can challenge and test the ongoing hypothesis, a critical part of the research process. Pat and Tom both take advantage of my abilities to observe impartially, with no vested interest in either novel hypotheses or entrenched dogma. I neither see, nor know, more than anybody else (it is usually considerably less), but together we three challenge each other, and constantly apply the checks and balances that are needed to progress. We all know the tendency to see what we expect or desire. I have seen this operate with lasting and vigorous conviction through long series of analyses. In art, it may be a creative gift. In religion, it can sustain you. In science, it can mislead you.

For the Mesozoic *Ausktribosphenos*, *Bishops* and *Teinolophos*, I was to employ the same technique and many of the processes I had used for rendering the much younger, giant marsupial *Diprotodon* (*see* Chapter 3).

My task in the case of these Cretaceous mammals was to 'visually' restore individual specimens. My greatest challenge was to cope with the scale of the miniscule material via a stereo microscope: the largest of these fragile, toothy slivers of bone was about 20 mm long, and the largest tooth was only 1.7 mm wide. This time, I set to work using a camera lucida attached to the microscope, and progressed to set my graphic measures so as to produce an orthographic outline. Accounting for the degrees of parallax that the microscope produced through the various magnifications was complex. I needed to resolve the details of the specimens despite the dimensional distortions of the stereo system. I also needed to eliminate the variation of the staining on each of the specimens, resulting from the petrification process, in order to clearly reveal their original physical form. As well, I was to 'repair' a myriad of fractures and misalignments – to retro-deform them. This was challenging, and to some extent subjective, and so I needed to proceed conservatively and consistently for each illustrated view.

Figure 12.15 *'Bishops – hypothetical posture studies'* (2000). Graphite on paper, 18 x 24 cm, collection of the artist (P. Trusler). Sketches produced in an effort to capture the furtive nature and possible appearance of an ausktribosphenid mammal for the 'polar winter hypothesis' art (figure 12.16). *Bishops* was the size of a modern day mouse.

Our growing awareness of more subtle geological changes became a point of continuing debate. The original features of the specimens recovered from these Cretaceous sites had been variably deformed owing to the fossils' plastic flow under pressure, as a consequence of having been buried to great depth and subsequently returned close to the surface during the episode when the present mountain ranges were formed. Such changes would have taken place no matter how large or small the specimens may have originally been. We knew that this variability was most likely owing to their orientation in the sediments with respect to the compressive forces that had operated over the intervening 100 million years or more. This was not always clearly expressed via detectable fractures. Caution was required: where samples are so few in number, there is reduced opportunity to assess the variations or consistencies than is provided when you have the luxury of comparing many specimens.

When I had produced the illustration of the brain cast of *Diprotodon*, I had chosen to represent the missing areas from my single specimen in a different tonal register. This was done to unambiguously show what was inferred through symmetry, clearly signalling how I had interpreted the form in these areas of loss with respect to the distortions that were evident. In this way, the distinction between fact and interpretation remained obvious to the viewer.

For the Mesozoic mammal jaws, I used the same device to indicate the missing parts of the specimen or any parts of the restoration that could not be unambiguously resolved, but did not reconstruct a complete jaw. We really had no means of prediction at the time I prepared the reconstruction. When the drawings were published, they were accompanied by a clear photographic record to allow our reconstruction work to be examined throughout and our conclusions to be open for challenge.

The discoveries accumulated and, with the increasing number of specimens, the analysis of them could be refined. Greater risks could be taken with the preparation of the delicate fossils once we had several of a given type. In turn, this gave clearer insights and led to revised thoughts, so the artwork, in some cases, could be refined too.

Despite the 'payload' of mammal remains, to this day they are almost entirely jaws! Curious but true, and yet these were providing Tom with a variety of forms – some that showed close relationships and others that were quite disparate. From this one point of entry, the scientists could begin to propose that there was probably a relatively numerous and diverse small-mammal fauna in this part of Gondwana at that time. Curiosity was certainly nibbling away at our thoughts. While Tom and his colleagues argued the identity and meaning of the finds (the scientific ramifications of which were quite profound), I think the entire excavation and research team wondered about the life of these diminutive furry denizens of the Early Cretaceous forests as they most likely scurried about the feet of dinosaurs.

We had no direct way of building an image of them; no way to put a face on any ausktribosphenid! The prospect for reconstructing such a diminutive monotreme as *Teinolophos* was also remote because, even though we had some 'living fossils' in the modern echidna and platypus, these were so very highly derived and divergent from each other as to be of very limited comparative value for reconstruction. Intermediary fossils of sufficient age were few and fragmentary, and certainly not well enough known to begin to track monotreme relationships for my purposes. However, the significance of the new ausktribosphenids demanded something be done to at least provide a visual background to the unusual story that was unfolding. A concept illustration was needed, in the manner of the earlier 'Aurora' painting (*see* Chapter 7), to place these in context with the environmental hypothesis. *Bishops* was the ausktribosphenid genus represented by the best specimens and Tom had a clear understanding of the evolutionary significance of the group from the features expressed by the jaws and teeth. In a sense, he was as surprised as anyone by the morphology, but he knew what they were not – neither marsupial nor monotreme, to put it simply – and asked me how I could convey such negative thoughts in an image.

I had no idea. I told him that it was going to be subtle, but I could start creatively sketching, testing my skills, and he could guide my progress. Gradually my little mammal images began to align with his thoughts – guided as much by our mutual understanding of the modern mammal groups that we did not want it to look like. The intermediary features of these jaws broadly suggested the relative position of the group with respect to some other Mesozoic mammals, and we had some generally held consensus from the reconstructions of other better-known fossil groups to guide us as well. There was some measure of control to our speculation! Above all, we wanted to bring people into the possibilities of this Mesozoic world: that some of the profound evolutionary developments of the time were at play on a much smaller stage. We hoped that the design of the resulting painting would convey just that.

(Following pages) **Figure 12.16** '*Bishops*' (2000): reconstructed scene of the Early Cretaceous 'polar winter' of southeastern Australia. Alkyd oil on linen, 30.5 x 43 cm, National Geographic collection (P. Trusler). A hypothetical winter scene in polar southeastern Australia 115 million years ago. The tiny mammal *Bishops whitmorei* peers out at a group of the ornithopod dinosaur *Qantassaurus intrepidus* as they make their way along the shore of a frozen lake. Peter produced this painting after Pat and Tom learned they were to receive the Chairman of the Committee for Research & Exploration of the National Geographic Society Award, 2000. The painting was presented to the committee at the ceremony where Pat and Tom received the award. Barry Bishop, the former chairman of the committee, and Frank Whitmore, the former vice-chairman, supported the work that led to the discovery of the species named in their honour for many years – during which very modest results were obtained – in the conviction that if they did so long enough a significant favourable result would occur, which it eventually did.

13
The Last of the Mob

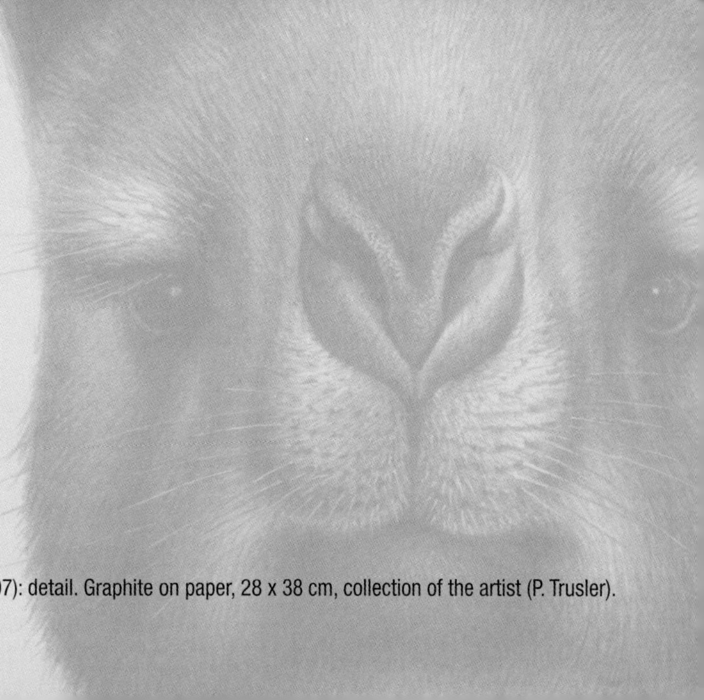

Figure 13.1 '*Procoptodon goliah* – life reconstruction of head' (2007): detail. Graphite on paper, 28 x 38 cm, collection of the artist (P. Trusler).

When I was a kid, I loved to collect stamps. This interest continued into adulthood, but I restricted my gatherings. As my studies narrowed to palaeoornithology, I zeroed in on bird and fossil stamps, and I still have a file in my professional office that reads 'Bird Stamps'. So I was rather chuffed when Rosemary Clark from Australia Post called and made an appointment to see me at Monash. This was in the early 1990s and it was only later that I realised her call was the beginning of a long, and quite wonderful, association with Australia Post.

Rosemary came to talk to me about putting together an Australia Post stamp issue on Australian dinosaurs. After we had chatted for a short time, she asked my advice concerning who could help her develop the concept and provide the scientific data for such a stamp issue. My eyes brightened and a big smile must have spread across my face as I said something like: 'You've come to the right person!' She was delighted, being the head of a team of philatelic researchers whose job it was to source information, design the issue and commission the art. She certainly had come to the right place, for at Monash and Museum Victoria, she and her group had everything they needed concerning the science. As well, Tom and I had long been working with Peter to compile the images that were quite relevant to this Australia Post project.

Unbeknownst to me, Rosemary must have been quite aware that this was an auspicious time to be producing such a stamp issue, for in 1993 there was to be the launch of a new Spielberg movie, *Jurassic Park*, which would grip the world audience more than any other dinosaur film had ever done. The stamps were to be launched in the same year, reinforcing the fascination that the public has always had for these strange and ancient reptiles, the dinosaurs. I was busy working with the Russians on launching the *Great Russian Dinosaurs* exhibition, only the second large dinosaur exhibition to visit to Australia. In the end, the stamps were issued on 1 October 1993, the *Great Russian Dinosaurs* opened in August of that year and *Jurassic Park* had sellout crowds in movie theatres across Australia – all nearly simultaneously.

I can remember taking part in the press preview of *Jurassic Park*, and using this opportunity to note to the journalist in attendance what seemed outwardly to be a planned coordination of the dinosaur stamps and the launch of the dinosaur exhibition. My team and I at the Monash Science Centre had even put together a modest exhibition at the Australia Post headquarters on Exhibition Street, a dinosaur display that we linked to Museum Victoria's *Great Russian Dinosaurs*, on show just up the road on Russell Street. On the first weekend the *Great Russian Dinosaurs* exhibition was open, more than 11,000 visitors attended. I was once told by the Philatelic Group that the dinosaur stamp issue was the second largest in the history of Australia Post, and if all the stamps were laid out on the ground, they would have covered more than two hectares – this was a very large stamp issue! I had made sure that not only were there stamps but also educational materials that school kids could use, and the Philatelic Group had developed a number of beautiful postcards and accompanying paraphernalia to accompany this special issue for children. All the funding for Peter's artwork was borne by Australia Post, and my fledgling Monash Science Centre was provided with funds to develop the educational material and for me to provide the scientific text and expertise. In 1993 the $5000 grant provided to the MSC was quite significant in our budget and thoroughly appreciated.

I had written to *Time* magazine in the hope that they might find the interrelationship of all these projects of interest. I also indicated to them that a fair bit of new art had been rendered for this project and that there was also a sizeable international exhibition about to be staged, of Russian material. This caught their attention and Graham O'Neill, their head science writer, took the lead on this combination of stories. Graham's work headlined Pacific editions of *Time* in August 1993 with two lengthy articles – perfect timing to highlight each of these interconnected projects. I treasured this moment: it had taken some time to bring it all to fruition.

A person critical to helping me put all of the pieces of the puzzle together was Kay Hamilton, a public relations specialist. It was she who introduced me to the Channel 9 management, who in turn introduced us to the Hambleton Ruff group in Melbourne. Neil Ruff and Rick Hambleton then introduced us to senior management in Qantas Freight. These were the funders, publicists and movers that truly made it possible to bring the *Great Russian Dinosaurs* to Australia. And it would later be Kay who made further connections with Australia Post, when I was making my bid to do a further stamp issue, *Creatures of the Slime*, on the Precambrian record of Australia. All involved in these multifaceted projects were in it as much for the chance to do something really different and unique as they were for profits and public relations. As an academic, it was most enlightening for me to see the same inner fire that drives our curiosity alive and well in the business world. I have struck this idealistic approach many times since, most recently in my dealings with ConocoPhillips in my educational and exhibition work in Timor-Leste. There is a genuine spirit of building infrastructure and advancing education in some managers of business, and to me this was truly inspiring. Real profit comes when there is more to life than money, though, of course, one must manage the dollars as well!

When this first satisfying project with Australia Post began, in the back of my mind – as a palaeontologist and the co-author of many books and research papers on the fossil history of Australia – lurked a cunning plan not to just finish with the dinosaurs. In the longer term, I hoped to see another few series of stamps that covered the rest of the geologic time scale. We had just 'constructed' a snapshot of the Mesozoic Era – our image of this continent about 110 million years ago. I hoped that in the future I could facilitate stamp issues on the older Palaeozoic Era and the younger Cenozoic Era. Later, when I began studying the Ediacaran, I truly hoped for an issue featuring that period of time too. I did mention this to Rosemary, and she was very encouraging.

(Previous page) **Figure 13.2** 'Australian megafauna' (2008): philatelic panoramic illustration featuring, top left to right: *Genyornis newtoni, Diprotodon optatum, Procoptodon goliah*; bottom left to right: *Megalania prisca, Thylacoleo carnifex* and *Thylacinus cynocephalus*. Alkyd oil on linen over hardboard, 45.7 x 91.5 cm, Australia Post collection (P. Trusler). Peter's art on the megafauna of Australia found many presentations. Pat was delighted in finally managing to have a medal struck in honour of the art and these ancient denizens of Australia's past. It was Peter's *Diprotodon* image that was minted.

Time passed. I was distracted by other ventures. But when I finally managed to piece together funding for construction of the Monash Science Centre, and Building 74 on the Clayton campus was launched in 2002, I found myself with a bit of breathing space – better, bigger facilities and time to think. I could plan and refresh the education and exhibition programs and involved myself in a new field of research (*see* Chapter 11).

As the research progressed and work on *The Rise of Animals* was underway, Kay Hamilton and I approached the Philatelic Group at Australia Post about a second stamp issue. We knew that if we were successful there would be funding for the production of Peter's art. We were successful, and there began another most satisfying interaction with Australia Post, starting with the manager, Noel Leahy and in particular with Mary Hoban, in the planning of content and design. Again, Australia Post generously funded Peter's work, as well as funding me and my Monash Science Centre staff to write the content of many associated educational products that were put together for school kids and the public.

The delightful finale to all this was our participation in the Pacific Science Explorer Expo in 2005, where literally tens of thousands of people from around the world attended to meet with postal organisations from many parts of the world. There were hundreds of thousands of stamps and stamp-related products on display. This sort of huge event was held every four years somewhere in the Pacific region – this time in Sydney. It was gratifying for those of us involved – including Peter Trusler, Jim Gehling from the South Australian Museum and Andrew Plant who had written an accompanying book on aspects of the Precambrian with a little help from me. Australia Post had asked me to organise some events for children attending this expo, and two of my staff, Lesley and Gerry Kool, set up a mock palaeontology laboratory where visitors could make their own casts of some of the very Precambrian animals spotlighted on the stamps, from the moulds we had made from the original specimens in the South Australian Museum collections. This temporary lab was in operation for five days and was simply fun; it would be hard to estimate how many thousands of plaster casts went out the door in the hands of newly inspired palaeontologists. Peter's art and the lab activities certainly were opening the eyes of visitors to a world most of them had never known. And even more pleasing, our stamps sold like hotcakes, and were some of the most sought-after in the entire expo.

That was 2005. I still had the longer-term plan in my head, but other things intervened. My work on the Precambrian became intense, in particular my taking on significant duties as one of the co-leaders of the International Geological Correlation

Project 493 (www.geosci.monash.edu.au/precsite). My part in organising two of the three international conferences on Neoproterozoic animals took time, as did being the principal writer or editor of two books – one for the general public and the other a professional compilation of the papers from the two conferences, one in Prato (Italy) and the other in Kyoto (Japan). I had little time beyond carrying out my job as director of the Monash Science Centre and maintaining my teaching and supervisory roles in the School of Geosciences at Monash to think about much else, bar getting the occasional bit of sleep.

Eventually, however, I received a call from Mary Hoban concerning the possibility of taking on another stamp issue, this time on the recently extinct megafauna of Australia – really 'the last of the mob' of our enigmatic fossil past. This sounded absolutely according to plan, and certainly was magnificently initiated this time by Australia Post. Over a delicious lunch with Noel Leahy, Mary and part of the Philatelic Group, our plans were put into play.

Of course, from the beginning, it was agreed that Peter was to render the art. Again, I was delighted with not having to come up with the funding, either personally or by writing numerous grant proposals that more often than not ended in failure after considerable effort had been put into them. Australia Post, heroically to me, underwrote not only Peter's art but also the writing of the text for a number of productions by this governmental body – truly significant to me and my colleagues both in the School of Geosciences and the Monash Science Centre, who are dedicated to educating young kids, and the general public, about science and to tweaking their curiosity about the natural world in which they live. It is a world in trouble for its very future existence and needs ever-increasing nurturing.

Much of the work I had carried out for my own PhD on the evolution of the environments occupied by my simultaneously evolving giant ground birds, the dromornithids, formed a part of the background Peter needed for his reconstruction. He already had his extensive database and experience gained in his preparation for the Bacchus Marsh *Diprotodon* project, *The Fossil Book*, *Wildlife of Gondwana*, *Magnificent Mihirungs*, as well as his observations of the natural world in his travels around Australia over decades, and his even earlier works on *Birds of Australian Gardens* and the Simpson and Day book on living Australian birds. He had been in the field with both Tom and me, and all of us had engaged in hours and hours of discussion of things palaeontologic. This was a formidable beginning.

But, in Peter's meticulous style, he had much more that he wanted to look over – and so he did. He discussed the latest findings with other researchers, such as Rod Wells at Flinders University with respect to the marsupial lion *Thylacoleo*, and Peter Murray and Tom about *Diprotodon* and its relatives, the diprotodontids. Peter Murray had significant insights about the dromornithids as well. Just around the time we began working on the megafauna stamp issue, a set of footprints, possibly those of *Diprotodon*, was discovered in the dry bed of one of Victoria's Western District lakes. Peter went along with Dave Pickering – preparator and long-time digger with Tom, Lesley Kool and our crews along the Victorian coast – to examine these prints, just so he could get the feet of *Diprotodon* in the stamp image right; it was not the standard way these feet have been painted in the past.

With the imagination working overtime in the Philatelic Group, I received many, many emails about new text needed, and again I was absolutely over the moon. Besides the usual items produced to accompany the stamps – and there were even more this time, special postcards and a showy folder with the entire stamp panel and the individual stamps, and a variety of presentations of the first-day covers – was to be a checkbook filled with little booklets of 10 x 55c stamps. Additionally, Peter and I put together a fact book with his sketches and graphic overlays rendered with his own research findings, along with input from a number of researchers (Rod Wells, Peter Murray, Aaron Camens, and several staff from Museum Victoria, South Australian Museum and Monash University), and designed beautifully by Adam Crapp, one of the philatelic staff.

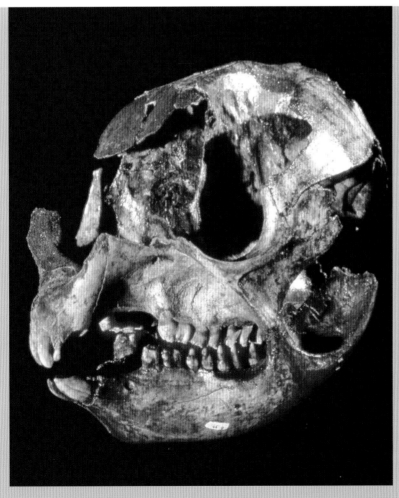

Figure 13.3 Pleistocene cave deposits have yielded well preserved remains of some Australian megafauna species. *Procoptodon goliah* was a massive kangaroo, and this skull from Victoria Cave, South Australia provided unique insights for the reconstruction illustration featured in the Australia Post philatelic issue. The lower jaw is about 30 cm in length (photo by S. Morton, courtesy of Rod Wells, Flinders University).

To my further delight, the Philatelic Group asked me to write a short book about all of the stamp issues so far that had dealt with palaeontology, and so I was able to concisely discuss the meaning of each of the 'time pieces' (Precambrian, Mesozoic, Cenozoic) that I had cunningly planned long ago. My Monash Science Centre education staff, headed by Michael Roberts, helped produce the *Megafauna: Facts and Fun for Students* handout sheets, making sure the exercises and graphics alongside Peter's art were compliant with the Victorian State curriculum standards, and so useful to teachers.

The pride of place, however, goes to the production of a dream I had from the beginning of my work with Australia Post – to have an official medal minted. This was done for the megafauna, using Peter's concept of *Diprotodon* as its principal image. My single copies sits, most treasured, at the base of my research computer at Monash, just another way of educating the curious public, I think.

The megafauna stamps were launched at the Monash Science Centre to coincide with the opening of our new *Wildlife of Gondwana* exhibition by Stephanie Fahey, our deputy vice chancellor, international, and Noel Leahy, the head of Australia Post Philatelic. Both displays featured Peter's art from many projects – art that was selectively placed beside the actual bones and teeth they were based on in the exhibition. And, for the first time in the history of Australia Post, a university post office (the Monash University post office) was designated the issue point – first day of issue 1 October 2008!

This third issue was titled 'The Last of the Mob'. My thinking at the opening was that I hoped that it was not the last stamp issue on prehistoric life – for we still had the Palaeozoic 'mob' to go. Then my plan to image, in stamps, the main blocks of the geologic time scale and its biota would be complete. But that will be another story!

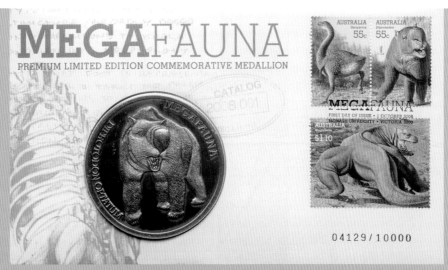

Figure 13.4 A special medal minted on the occasion of the release of 'The Last of the Mob' stamps (S. Morton and courtesy of Australia Post).

The Artist: Peter

When it comes to the popularisation of science, and particularly when using palaeontology as a tool to stimulate enquiring minds, dinosaurs will always be the obvious choice with children. The intrinsic character of dinosaurs seems to fit well with cultural concepts of fearful and dangerous beasts. Dinosaurs provide a safe fantasy in which children can indulge their developing emotional and intellectual sensitivities and on which they can build a sense of their place in this world. The psychological processes at play here are for the great part similar to those at work in the realms of populist fantasy writing, classic sagas and traditional folk tales.

For educators, dinosaurs therefore provide an easy beginning. Once enthusiasm has been harnessed and enquiry kindled, it takes little time and effort to broaden and deepen a natural curiosity and branch out into other subjects. Because palaeontology requires such a multidisciplinary approach, there is inherently a broad spectrum of relevant and fascinating avenues that may be explored. The field enables a great degree of flexibility for introducing other realms and quite complex concepts.

Having successfully introduced such an esoteric subject as dinosaurs for a national philatelic issue, an obvious follow-up was going to be the *Megafauna* issue. In the context of the contemporary purpose, this too, might seem obvious. However, I say 'esoteric' because I reflect on the traditional subject matter in philately – this being primarily royalty, heads of state, national icons, military heroes, etc. The vast majority were dead people! Australia was probably among the first to include native fauna on stamps, if only for the young nation's lack of suitable historically significant personalities. Perhaps a national identity crisis, which could only find immediate resolution by looking to the unique inhabitants of its vast natural landscape? The concept of national identity as expressed through the graphic iconography of legal tender – banknotes, coins and stamps – is not something that I had deeply contemplated. The process and search has widened in all respects over recent decades and the fact that illustration in this sphere now includes the prehistoric is perhaps a sign that there is a widely felt cultural awareness of the rapidly shrinking place of humans in the workings of the planet's history. Ironically, it is coming with the recognition of the profound effects that we are inflicting on our planet. National has indeed gone global! Are we again at a crisis and looking to the dead inhabitants of this planet for meaning and identity?

The instigation of this philatelic concept involving a series of stamps within a wider pictorial sheet had been deliberately aimed towards early education and stimulating a wider interest in stamp collecting among a young audience – education and economics. Following the success of *Australia's Dinosaur Era* mini-sheet issue, the large philatelic format has featured annually as an integral part of Australia Post issues. The subjects have been diverse. Over this intervening time, the Philatelic Group at Australia Post and our team had pulled off an even greater coup: the 2005 issue concerning the origin of Precambrian animal life – *Creatures of the Slime* – was a far more unusual and remote subject to deal with in terms of the presentation of scientific understanding to a broader public.

The megafauna constitutes a recent biota, the largest representatives of a modern faunal component that has almost entirely vanished. Yet so many were within our reach. Those species that have remained, most notably in Africa and parts of southern Asia, are currently suffering desperately at the hands of humans. We all know the surviving megafauna so well. They are the classic zoo animals that populate our childhood memories and still fuel our wonder at the natural world – elephants, lions, rhinoceros, and giraffes. These creatures, too, have been the objects of exploitation, and even now it seems we are 'loving them to death'.

I was amazed when I first realised that these magnificent beasts were but a small fraction of the megafaunal menagerie that had once roamed all continents (with the exception of Antarctica) until the late Pleistocene – even more amazed that it was less than a few thousand years ago in isolated cases! The accelerating cycle of climatic events of the Pleistocene, which spurred their radiation and their attainment of gargantuan size, eventually diminished their variety and contributed to their demise.

But, perhaps climate was not the only perpetrator – prehistoric humans progressively encountered the megafauna as they spread across the globe. Coincidentally, it seems that the biggest beasts vanished in this process. On some continents, the addition of humans to the potential list of multi-causal factors contributing to the megafauna's plight was the final tragedy. The evidence of interaction between humans and megafauna is provided among the earliest cultural markings – cave art, engraved bones and artefacts, both practical and decorative, such as bone and ivory tools from mammoths and cave bear claw necklaces. Giant bones bear the scars of butchering and burning in ancient hearths, testifying to human consumption. On continents such as Australia the coincidence of events is something of a smoking gun, because precise evidence of humankind's direct role in the extinction process is scant. Just what or who pulled the trigger is fiercely debated. The impact of Europeans on natural Australia as we understand it to have been at the time of their arrival, has been catastrophically rapid and dramatic. I personally find it tantalising to think that we have just missed the megafauna in this part of the world, because I now realise that I may have grown to love a 'natural Australia' that had already been highly modified since the first wave of human occupation about 40 000 to 50 000 years ago. My frame of reference has shifted.

Palaeontology has taught me that so few people are privileged to fully comprehend these recent changes. It seems to me that the all-important issue of rapid climate change needs to be understood and appreciated, and if that is to be accomplished, we would be well advised to communicate what we do know about the story of the world's megafauna. It is a vivid example of what can happen, because it actually did happen – only yesterday!

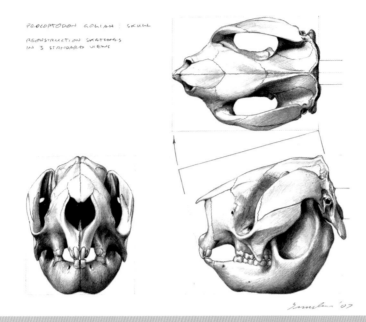

Figures 13.5 '*Procoptodon goliah* – skull reconstruction' (2007): three standard views. Graphite on paper, 28 x 38 cm, collection of the artist (P. Trusler). Figs 13.5 to 13.9 illustrate the design sequence for the giant, short-faced kangaroo, *Procoptodon goliah*: studies of the brachycephalic head and distinctive single-toed skeleton of this now extinct form. The lower jaw measures nearly 30 cm in length.

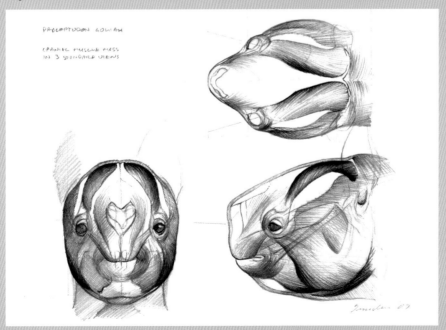

Figure 13.6 '*Procoptodon goliah* – muscle reconstruction of head' (2007): three standard views. Graphite on paper, 28 x 38 cm, collection of the artist (P. Trusler).

Figure 13.7 '*Procoptodon goliah* – life reconstruction of head' (2007): two standard views. Graphite on paper, 28 x 38 cm, collection of the artist (P. Trusler).

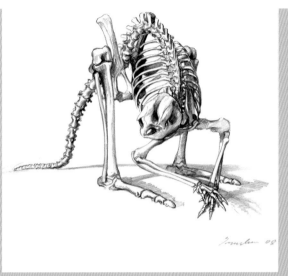

Figure 13.8 '*Procoptodon goliah* – skeletal design' (2007). Graphite on paper, 22 x 28 cm, Australia Post collection (P. Trusler).

Figure 13.9 '*Procoptodon goliah* – life reconstruction design' (2007). Graphite on paper, 22 x 28 cm, Australia Post collection (P. Trusler).

I feel that our contemporary world is something of a fool's paradise – one with a poor concept of time, an un-empathetic concept of loss beyond our immediate tribe and little willingness to exercise our developed intellect outside of the human hunger for social status. All of this is ably supported by a brilliant, voracious and complex technology and the exploding population that is occurring because of our temporary success. We are not at all immune from the natural flux of biological processes, and both science and history can easily tell us how these events have invariably played out in the past. They are likely to continue into the future – a future in which our children, and their children, will be players.

With these absolutely depressing thoughts in my head, it became so important to me that the fruits of hard-won labours in biological and behavioural science be heard and comprehended clearly. Yes, those dreamers and thinkers who pursue such seemingly esoteric interests have something profound to offer to the human ego. At the same time, I can clearly appreciate that the precision of our knowledge needs considerable refinement. The debate about the detail of the processes at play needs to continue with vigour. For our part – artists and scientists alike – the trends in this process are nevertheless very profound and should not be ignored because of lack of universal academic consensus.

The crisis, while sobering, also intrigues me. It not only presents an urgent need for us to reconnect with the natural world in a way that is considerate of it, but it also can be the vehicle through which we can attain a greater understanding of our place in this universe and of ourselves. Just as my painting of natural subjects fostered my interest and youthful joy, I hope this crisis may sharpen our focus on the natural world so that it may continue to be the wellspring for others.

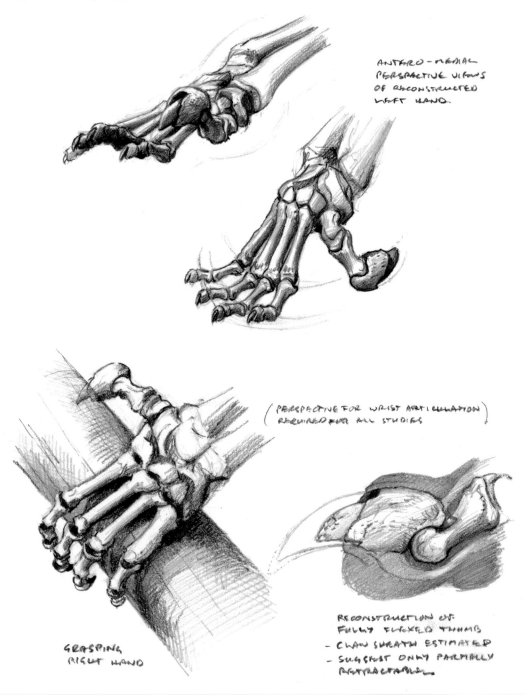

Figure 13.10 '*Thylacoleo carnifex* – manus skeletal studies' (2007). Graphite on paper, 38 x 28 cm, collection of the artist (P. Trusler). The 'hand' of a near-complete specimen of *Thylacoleo* from the Nullarbor Caves, Australia.

With this in mind I returned to stamp illustration, taking on the commission to feature Australia's megafauna with renewed vigour. I had a brief to represent a small vision of the Pleistocene landscape from the 'Aussie Outback'. This just might communicate concepts of change and of the ecological relationships between a variety of megafaunal animals to the public – to children especially. We sought to introduce a greater awareness of the identity of six megafaunal 'characters'. This was to be another panorama, such as I had previously rendered for the age of dinosaurs, but this hypothetical landscape was to be populated by some of the most characteristic and better-known representatives from this time, somewhere between 50 000 and 100 000 years ago.

A round-table discussion ensued between the philatelic staff, my scientific advisors and myself, and six species were chosen – a big bird, a giant reptile, and four unusual marsupials. The project began with the basic constraints imposed by the composite philatelic format, but other than this, I had free reign of all other aspects.

The panoramic genre is extensively used in palaeontological art. It retains considerable power, despite my reservations about limits to its aesthetic and scientific value.

This strength is partly due to its long tradition and accepted worth in both science and public perception. The concept is one of pragmatism and allows for the presentation of a diversity of information and multiple narratives. Interactions between all the elements can easily accommodate several different conceptual ideas and processes in one image. However, there are limits to the complexity that can be illustrated in this way.

There is an interesting and very fundamental dichotomy operating here that goes to the core of my perceptual training. It may have a cognitive origin, and it certainly varies between people. It could be somewhat crudely expressed as image versus language or, vision as distinct from thought. I may have had an aptitude or predilection for one over the other, but it certainly took years of perceptual training to develop my visual logic. As my painting progressed, it became apparent that I needed to overcome my lack of fluency with the language mode as it applies to art practice. I ultimately realised that these two different mental processes of perception were not mutually exclusive.

This panoramic genre can be a victim of its own artifice in this respect and its success greatly depends on its acceptance by the viewer. It constantly carries the risks of contradiction, generalisation, stereotyping and cliché. This not only tests the creator's ability to process a myriad of external influences at the outset but also ultimately places the artist at the behest of the viewer. This can limit the presentation of novel thought – one reason for my reticence in using this style. I've certainly found that such designs test my aesthetic and practical skills in the challenge to overcome the inherent shortcomings of this form of presentation. When applying detailed realism to panoramic composite illustration, the reality of the total effect is critically dependent on artistic 'trickery' – illusions.

It is interesting that one should need cunning to present a series of facts. On one hand, I am endeavouring to make something visibly believable that is scientifically supported, and on the other, asking the viewer to suspend their disbelief. For example,

Figure 13.11 '*Thylacoleo carnifex* – philatelic design' (2007) and resulting stamp. Graphite on paper, 28 x 22 cm, Australia Post collection (P. Trusler). Design sequence rendered to establish the appearance and behaviour of *Thylacoleo carnifex* (Figs 13.11 to 13.15).

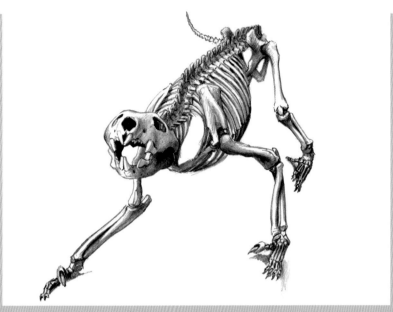

Figure 13.12 '*Thylacoleo carnifex* – skeletal design' (2007). Graphite on paper, 28 x 22 cm, Australia Post collection (P. Trusler).

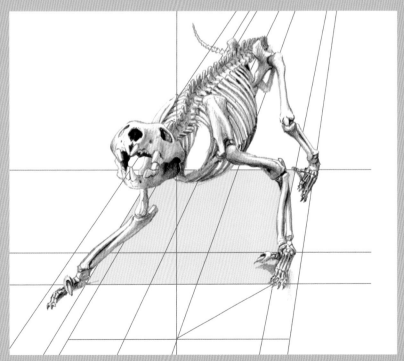

Figure 13.13 '*Thylacoleo carnifex* – skeletal design' (2007): including digital perspective overlay. Graphite on paper, 28 x 22 cm, Australia Post collection (P. Trusler).

Figure 13.14 '*Thylacoleo carnifex* – muscular design' (2007). Graphite on paper, 28 x 22 cm, Australia Post collection (P. Trusler).

Figure 13.15 '*Thylacoleo carnifex* – life reconstruction design' (2007). Graphite on paper, 28 x 22 cm, Australia Post collection (P. Trusler).

the narratives of the *Megafauna* panorama are based on sound understanding, and yet the juxtaposition of elements in the design is both spatially and temporally unlikely. In other words, there is an inconsistency or unevenness to the visual logic of such an image, but not necessarily the scientific narratives.

Stepping aside from the science for a moment, this scenario is nothing new to art. The absurdly complex battle scenes of the 18th and 19th centuries, or the tumultuously erotic heavens of the Italian high Baroque, immediately spring to mind. The digital effects and animation of the contemporary motion picture industry can now play the same unfettered game with grandiose narratives in visual form. Being 'epic' is not my forte, and so this everything-at-once visual storytelling runs a little counter-current to my sensibilities – having matured, as I did, with the reductionism of the modern art aesthetic.

Children respond to the conundrum without being totally aware of its implications – they simply revel in the daring spectacle. Pat, Tom and I have felt that such imagery is valid for the communication of a range of scientific ideas generated through palaeontology, especially for these philatelic products and for *The Fossil Book*.

And so, I went back to immerse myself in the museum collections of megafauna bones. There was some catching up to be done with old acquaintances such as *Megalania*, *Diprotodon* and *Genyornis*. There were many more fascinating discussions to be had with palaeontologists, such as John Long, Trevor Worthy and Rod Wells, to name a few. Piles of books and scientific papers had to be read and conferences needed to be attended. I was invited to view new discoveries and to assist in illustrating them. Rod Wells introduced me to his student Aaron Camens, for he, together with Stephen Carey and Dave Pickering, was working on fossilised megafauna track-ways. The variety of recent evidence was refining the story, and all helped me to build up a clearer picture of the animals and their world.

For the design, the megafauna's world was to be centred on a drying waterhole, thereby providing for the interactions between the varied species that would have been compelled to congregate there. The habitat structure and many of the plant genera would be more familiar to us from the coastal ranges of contemporary plant communities. The proportional representations of each may have shifted in recent times and the central deserts would have supported larger areas of mature woodlands than they do today. The prediction of such matters is not precise for my part; I have often contemplated the visual effect of the feeding pressures that the herbivorous members of the megafauna would have had on the habitat structure. The pruning effect of the modern African grazers, and especially the browsers, can be dramatic. A large proportion of the megafauna were browsing species, and thus the ecological balance that existed between them and their fodder has gone. Perhaps the loss of the plant species on which many depended was another likely factor in the cycle of change.

The anatomical reconstructions for each animal entailed the usual issue of body bulk and size. The proportions of such large and unfamiliar animals instantly look implausible to our eyes. The ultimate size of the reproduced images was another factor, but I had coped with this before. Inventing an irregular topography for the waterhole provided me with the option of varying the low-angled linear perspective with various

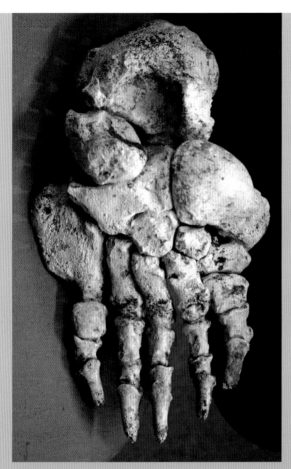

Figure 13.16 *Diprotodon optatum*, articulated fore-foot from Lake Callabonna, South Australia (photo by P. Trusler, Australian Museum collection).

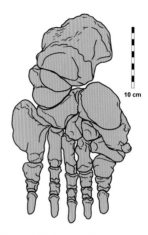

Figure 13.17 Graphic of *Diprotodon* fore-foot (P. Trusler).

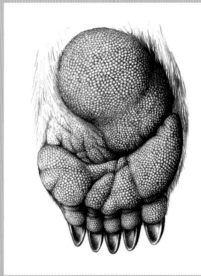

Figure 13.18 '*Diprotodon optatum* – life reconstruction of fore-foot' (2008). Graphite on paper, 28 x 22 cm, collection of the artist (P. Trusler). Reconstruction of foot pad structure of *Diprotodon*, based partly on the skeletal anatomy and comparative myology. The detail of the papillae on the surface of the *Diprotodon* palm has been based on the track-way cast of *Eowenia*, its ancestor.

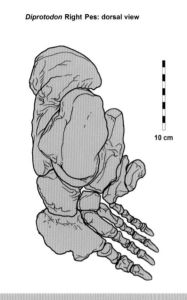

Figure 13.19 Graphic of *Diprotodon* hind-foot (P. Trusler).

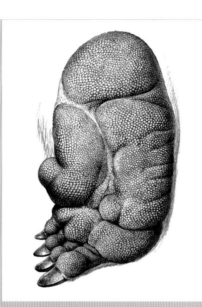

Figure 13.20 '*Diprotodon optatum* – life reconstruction of hind-foot' (2008). Graphite on paper, 28 x 22 cm, collection of the artist (P. Trusler).

Figure 13.21 The preserved detail of the diprotodontid track-way shows how the marsupial's hind-foot pressed into the muddy volcanic ash and partially obscured the print of its fore-foot as it walked. Scale in cm (photo by A. Camens, Adelaide University).

elevated surfaces that enabled me to bring some species to the fore. Having selected a suite of larger animals, I was missing sufficient smaller comparisons of scale to impress the viewer with the size of the megafauna. I therefore included familiar birds suited to the environment and time to assist the solution. Most importantly, the thylacine was to be included – the Tasmanian Tiger was a national icon, synonymous with the public perception of extinction. The thylacine's famously convergent wolf-like form, tiger-stripes and familiar body size were going to be crucial to imparting both the scale of the scene and its relatively recent time frame.

Figure 13.22 Pleistocene track-way of a diprotodontid preserved in volcanic ash sediments on the floor of a dry lakebed in western Victoria (photo by P. Trusler).

Figure 13.23 Australia Post 'Megafauna' stamps mini-sheet.

The last of the captive thylacines had died in the mid 1930s. Validated sightings have long ceased, although the folklore about the animal's survival has not entirely subsided. The thylacine was once widespread throughout the Australian mainland. It was part of the megafaunal ecosystem and it survived the major extinction event at around 50 000 years ago. It shared the landscape with the new human arrivals and was beautifully depicted in the earlier rock art. The introduction of the dingo to the continent some 6000 years ago changed the fortunes of the mainland thylacines, and they were quickly driven to extinction by the competitive placentals. The land bridge that had connected Tasmania to the mainland during glacial maxima had gone by this time, and so the island became the last refuge for the thylacine. European arrival was the final blow for the species and in a little over 100 years the tigers had vanished. That is expressing it mildly, for during that short period they witnessed a multi-pronged ecological assault: there was massive land clearing, logging and the introduction of domestic dogs; their eradication was promoted via a government bounty; they were collected for zoos and, while they were collected and studied for science, their plight was largely ignored by the professional community of the day. In this respect, the Thylacine was the only member of my visual panoramic narrative for which the extinction process was actually documented. The general knowledge about this animal ensured that it would provide immediate accessibility to the substance of my illustration – contemplation of the story of Australia's megafauna.

The thylacine was truly 'the last of the mob'.

Figure 13.24 '*Diprotodon optatum* – skeletal design' (2007). Graphite on paper, 28 x 22 cm, Australia Post collection (P. Trusler). Articulated skeletal design in perspective for *Megafauna* panorama illustration.

Figure 13.25 'Diprotodon locomotion study' (2008): postures based on track-way site, Camperdown, western Victoria. Graphite on paper, 14 x 35 cm, private collection (P. Trusler). Exit Diprotodon.

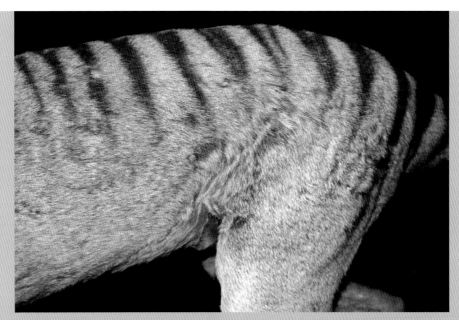

Figure 13.26 *Thylacinus cyanocephalus*: the stripes of the tiger. A mounted skin of the famous Tasmanian marsupial carnivore from the Western Australian Museum collection (photo by P. Trusler).

Figure 13.27 '*Thylacinus cynocephalus* – philatelic design' (2007). Graphite on paper, 28 x 22 cm, Australia Post collection (P. Trusler). Posture study for philatelic design: skeletal studies and images of the last Tasmanian tigers in captivity in various zoos around the globe informed the artist in creating new postures for this illustration.

14. The Art of Humour and Enchantment

Figure 14.1 Inspired by a 1982 cartoon in the *Herald Sun* newspaper, Peter rendered this comment on the official naming of Dinosaur Cove.

The Scientist: Pat

Young minds need food. Lively graphic imagery, in particular cartoons, can provide this nourishment and quickly engage the youthful intellect, fuelling children's innate curiosity. The educational value of such mirthful art should not be underestimated.

> José Ramos-Horta, President of the Republic of Timor-Leste and co-author of the geology book for children, *O Mundo Perdido Timor-Leste (The Lost World of Timor-Leste)*

I sipped on my delicious, strong coffee, locally produced, while sitting in a thatch-covered, open air dining room at José's. It was seven in the morning. The smell of burning wood reminded me of my own childhood on a small ranch in California. Here in Timor-Leste, in 2008, the smoke came from the kitchens and heating fires of the houses nearby. In California, in my childhood, it came either from crop burn-off at the end of autumn or, more worryingly, from the wildfires that all too often destroyed the chaparral vegetation covering the hills and mountains to the east of our farm in the San Joaquin Valley. Here, sitting with Nobel Laureate and President of Timor-Leste, José Ramos-Horta, the scent was simply nostalgic.

José and I were discussing where I might be able to house an exhibition my friends and I had put together concerning the geologic history of the island of Timor, particularly of Timor-Leste, a country that had come into being in 1999 by vote of its people, under difficult conditions. I had been lucky to raise the funding to put the exhibition together and now was looking for the ideal place to set it up.

Our conversation ranged over many possible venues, as well as the content of the exhibition and the length of the fossil record. José suggested that we should now visit one of these possible sites, the Presidential Palace, which lay in the hills behind Dili. Preparations for this trip included organising the drivers and vehicles (five in all) – and the bodyguards. Not long before, José had nearly been assassinated right here in front of his traditional home by a rebel group, seemingly in an attempt to topple the government. Now he was carefully protected day and night.

The two of us, with our many companions, drove out of the heavily guarded front gate of his presidential compound, beyond the checkpoint manned by the Pakistani Army and onto the road towards Dili.

Along the way, up the side of the mountain and on through the busy Taibesi market, we continued to talk. First, José asked about the length of time for which we had a fossil record of life. We talked about the changing panorama of animals and plants over this long, long time. I presented José with part of my standard first-year geology lecture on plate tectonics. This truly pleased me, for this man had long been a true hero to me, and to have him interested in the science was a real treasure to me.

Eventually we reached the pink Presidential Palace, high on the hills, and inspected it. This was a beautiful place, but because it was so far from town it was not going to work for the school kids and public for whom the exhibition was meant.

I took some photos, and José then asked me where else I needed to go. We headed back to the headquarters of the Alola Foundation in central Dili, where I was working with Kirsty Sword Gusmao and her staff on developing children's literature in science written in the local language, Tetun. José had mentioned that he was writing some children's stories, and the idea popped into my head that here was an opportunity to draw many threads together. José and I could write a book for the kids on the geologic history of Timor-Leste, and that could be coordinated with the exhibition we were going to set up in Dili. That exhibition was to be presented to the government as part of the rebuilding of the National Museum, destroyed in 1999 when the country was born. The book would significantly enhance the impact of the exhibition, and could also be used to teach science in the schools, tying into the development of the national curriculum underway at that point. I suggested this to José somewhat later, and he was just as enthusiastic as me – so began the little children's book O Mundo Perdido Timor-Leste (*The Lost World of Timor-Leste*), which would result in two versions presented in four different languages!

Words are great, but for kids too many can be quite boring and off-putting. When I returned to Australia in late 2008, my solution to this problem was, of course, Peter Trusler! I floated the idea with Peter, and he too was intrigued. I then explained my concept: to use a popular Timor-Leste legend that almost every small child there knows – that of the boy and the crocodile. The legend goes something like this. Once upon a time there was a young boy and a baby crocodile. The baby crocodile was having a bit of a problem getting from his nest to the sea, until the boy came along and helped him to get there. Later, when both were a bit older they met again. After thinking about eating the boy – for he looked quite delicious – the crocodile consulted with his friends and decided that it was not such a good idea. After all, the boy had been critical to his being alive! So, they became friends and travelled all over the place – in the 'now'. My idea (approved by José) was to have the boy and the crocodile travel not in the now, but through time. I wanted them to travel back in geologic time, taking into account what my geologist buddies from the University of Western Australia, David Haig and Myra Keep, and I knew was recorded in the rocks we were studying in Timor-Leste. This meant a trip of some 250 million years!

Figure 14.2 José Ramos-Horta and children launching our book, *O Mundo Perdido Timor-Leste* (*The Lost World of Timor-Leste*), about the geologic history of this new island nation. Peter's art has enchanted all of the kids in Timor that have seen it, and the book is to become part of the science curriculum at the primary and early secondary levels for which it was written.

I wrote the original text in collaboration with José, and the manuscript went back and forth between us for a few weeks. When we settled on a first draft, I passed that to Peter, and he began his work – in this case the style was completely left up to him. He knew we needed a Permian landscape with sea lilies (crinoids) and brachiopods, bryozoans and trilobites, among other things. And he knew he needed to be familiar with the actual fossils that had been collected from the Timorese rock sequences. He also knew that the boy and the crocodile had to have some reality in their appearance, since they were to be the central characters in the book. As this was a book for primary children, it needed colour, and it needed enchantment. But, it also needed to be scientifically accurate, at least in the information about the fossils and the geology, no matter how fanciful the story.

I had a stack of research papers on Timorese rocks and fossils. With this material, Peter's drawings of the animals and background would be based on reality. As well as some field experience in Timor, I had fossils collected from different parts of the island and detailed photos of others. So the data needed by Peter were in front of us – and we had a small budget provided by my university and ConocoPhillips, an oil company currently drilling off the shores of Timor and keen on social support. Several people offered to translate; this was critical, as the book had to be in many languages if it was to be used – Tetun and Portuguese, and English beyond that for the future. Lotte Renault, a Brazilian with a PhD working for an NGO in Dili, and a specialist in children's educational literature, helped immensely with this project.

Figure 14.3 Front cover of the book, *O Mundo Perdido Timor-Leste* (*The Lost World of Timor-Leste*), written by President of Timor-Leste and Nobel Peace Prize recipient José Ramos-Horta and geologist Patricia Vickers-Rich. Artwork is by Peter Trusler; he and Draga Gelt (School of Geosciences, Monash University) designed the book.

Figure 14.4 A page out of *The Lost World of Timor-Leste*, showing a child standing atop a succession of rocks, from recent to Permian; the fossils of each time period represented in this region are enclosed in the layers of sediments, once deposited on the ocean floor. The purpose of the book was to show children how the long history of the Earth, and specifically of Timor-Leste, has left traces of the past.

Peter had the manuscript in hand about two months before it needed to go to the printer – not a great deal of time. The book had been scheduled for release at the same time as the opening by José Ramos-Horta of his new Executive Offices on 21 August 2009, and that date was fixed. It was most fortunate that we had a printer willing to work to a tight deadline, for our original plan changed rather dramatically right at the end of this project!

Once the draft of the boy and the crocodile book was completed, we had it reviewed. This was done in a number of ways: several colleagues were given copies, and I tried a preliminary version of it on the local kids in the highlands of Timor. The kids were delighted, enchanted especially by Peter's art. I listened to their comments and gauged their understanding and enjoyment (or lack thereof!) by the look in their eyes and the questions they asked. I took their suggestions on board.

When I returned to Dili after this fieldwork I learned that a colleague there had sent our book to a professional children's writer. Her comments were both helpful and surprising. She suggested some sequence changes in the storyline, and picked up a number of inconsistencies and facts that needed clarification. But she had some puzzling queries – one being that, since the crocodile was the authority figure in this book, how could the boy sometimes offer good advice? This troubled me, for one of the themes in this book was the idea that kids should have input and be part of conversations and decision-making, from a very young age; this had been important in my childhood and given me confidence in my own decision-making as I grew into adulthood.

Figure 14.5 When our draftswoman came down with Ross River Fever, to make her feel a bit better, Peter drew a cartoon of her taking for a walk the vector that had presented her with this ailment, a very large mosquito!

Figure 14.6 *Happy Birthday Old Timer*. Peter presented this to Tom for his birthday – with the jaws of the ausktribosphenids in the glass where most people with fake teeth would place their own. The humour reflects nearly 30 years of close cooperation between these two blokes and was rendered with a connection to their work.

Our reviewer also noted that we had not followed one of the 'rules' of writing for children: that the boy should be in every illustration. Peter and I had a bit of a giggle over this. In one illustration both the boy and the crocodile were obviously absent – the illustration being one of a scene in the Cretaceous of Australia, where a rather menacing theropod dinosaur dominates. The accompanying text went something like this: '"Let's get out of this time," shouted the Boy. "I do not want to get eaten!" So they quickly continued the time travel to nearer today.' Not a bad idea to stay out of this illustration! I guess, overall, my assessment of this reviewer's comments is that Peter and Tom and I have never really tried to fit into the standard mould, and certainly did not in this case. Our aim had always been, and was the case in this book, to communicate in our own style. And, if the ideas are truly communicated, it doesn't matter if our approach is 'the road less (or never) travelled'.

As we have worked together over the years, with many other colleagues and friends, another form of 'art' has reared its head – and that is humour. This has not only

Figure 14.7 'Hell's Dinos' cartoon used as advertisement for the exhibition *Dinosaurs of Darkness*, held in Rapid City, South Dakota at the same time as the annual national motorcycle rally (2001). Tens of thousands of enthusiasts descended on Rapid City for the rally – and the dinosaur exhibition, due to Peter's art being proudly displayed on a large banner in the town centre.

Figure 14.8 Logo that Peter rendered for the *Dinosaurs of Darkness* exhibition, which the Monash Science Centre has toured for years (*see* Chapter 7).

been a way of relieving tension when the deadlines impose stress, but also a very good way of teaching. Given the personalities of Peter, Tom and me, there is always humour even without the excuse of stress! And the result of this has been a whole series of cartoons – all rather personal and not produced to win any competitions.

One was generated at the time that Draga Gelt, our constant companion in drafting and design, was quite ill – she was unlucky enough to have contracted Ross River fever and clearly needed some cheering up. Peter sketched her walking (or being taken for a walk by) a very large mosquito – the very vector that had brought her the virulent gift! As a birthday gift for Tom, Peter once presented him with a drawing of an 'old-timer' with his false teeth in a glass beside him – except that the teeth belonged to one of the ausktribosphenids they had worked on together over many years!

Another cartoon resulted from my touring an exhibition on dinosaurs to the South Dakota School of Mines in Rapid City, South Dakota, which happened to coincide with a huge motorcycle rally held there each year. Peter, with a cheeky smile on his face, presented me with three tough-looking dinosaurs ('Hell's Dinos') on equally tough-looking bikes as a parting gift when I left to launch this exhibition. Needless to say the Rapid City PR folks thought this so appropriate that the cartoon went up on a large banner in the centre of town – and both the bikies and the palaeontologists got serious 'mileage' out of this. I doubt that so many bike-lovers have ever attended a dinosaur exhibition! The logo for that same exhibition, *Dinosaurs of Darkness*, was another piece of Peter's art, and certainly got across the concept of the title.

Related to Tom's and my work at Dinosaur Cove is one of Peter's classics – his reinterpretation of a cartoon that originally appeared in the *Herald Sun* newspaper. A man and his wife are lounging on the Otway beach, the husband looking out to sea. His comment is 'Dinosaur Cove, eh? Stupid name if you ask me!' – quite unaware of a large theropod preparing to devour him. Once having seen the cartoon, my students remember the name of this cove!

Other cartoon images of some of the first animals, the Ediacarans, playing in a shallow Neoproterozoic sea – *Dickinsonia* tossing *Tribrachidium* into the waves – are useful in getting across complex scientific concepts. In this case it was the idea of

Figure 14.9 Cartoon depicting the idea of global playgrounds for the Ediacaran fauna. The underlying concept was that, as the seas became less saline, these late Precambrian first animals were able to spread worldwide. Earlier, more isolated populations had developed in less saline conditions but could not spread further because they were unable to tolerate the higher salt content of the global oceans.

'global playgrounds' and the rapid movement of these early animals into seas that were becoming less saline. For the students to whom I lecture each year, such a cartoon image is a powerful way to communicate. Perhaps not everyone is as 'image sensitive' as I am, but I have found that many are. These images, and their associated concepts, stick in the mind where words fail. Certainly Peter's image of a Qantas plane decorated with several *Qantassaurus* (named in recognition of the airline's support for our research and education programs) has burned that dinosaur's name into my students' brains, even if they forget all the other scientific appellations!

Yet another cartoon image concerns the constraints imposed on his work by gaps in the fossil record. 'Palaeo Pete' rendered his explanation of why the image of *Tehuelchesaurus* lacked the tip of the tail and the detail on the skull.

Over the years, Peter, Tom and I have not tried to win any contest with these images. They have, however, been powerful tools of communication and, even more, a continuing source of delight, both to us and to many around us. And again, they carry with them not just the humour but also something relevant to either the events or the concepts they celebrate.

Figure 14.10 Peter's 'explanation' of why he did not include the detail of the head or the tail of *Tehuelchesaurus* in his 1998 reconstruction.

Conclusions

A work in progress. Reconstruction of a Cretaceous Ichthyosaur from western Queensland being rendered as we write. 'The Golden Ichthyosaur' (2010): reconstructed standard views of *Platypterygius australis* skull. Oil on linen, 71 x 96.5 cm, commissioned by Museum Victoria, in the collection of the artist (P. Trusler, courtesy of Museum Victoria).

An image has the capacity to communicate a vast array of detailed information instantly, with precision of both content and context. Images can evoke considerable emotion as well, but not necessarily with the same fidelity. The great advantage of image over text is one of time, of course – a lengthy description or relatively complex narrative can be summarised and so understood at a glance, whereas hours might be involved in understanding the written word. And, of course, vision is more universal than language.

Conversely, images fix information – they give it an immutable face, and this is not always desirable. Just as a movie might diminish the imaginative power of a novel, illustration cannot always hope to leave open a variety of interpretations or be tailored to embrace every imagining or understanding to the same extent as the written word. But, when coupled with words, appropriately-rendered visual images will greatly refine and enhance the dissemination of information from one person to another. In some measure, the best attributes of each descriptive modality can compensate for the limitations of the other. Most especially for our adventures in science, the processes and complexity of life on this planet can be communicated more easily using both.

If communication through image and words is to be of the highest quality, then time must be taken and every effort made to hone the precision of image and word. Such honing involves many things: careful research, precise and thoughtful rendering of the visual image and careful crafting of the words – all demanding a determined and ongoing development of skills.

The association and friendship between the three of us spans more than three decades. It began when we two scientists were on the hunt for an artist who had the skills that would allow precise scientific information to be conveyed in imagery alongside the detailed scientific descriptions of the ancient marsupial, *Diprotodon*. The condition of the specimens to be 'presented', to both the general public and to other researchers, was incomplete. What we needed was an artist able to synthesise images that combined these fragments and information from various disciplines. The images required a three-dimensional quality that drew observers in, communicating to them intimate details of anatomical form. But, more than just revealing the essence of the scientific descriptions, these images needed to challenge the viewer to ask more questions about the material on show.

We have, throughout our work together, continued to find each other's different perspectives and approaches informative. We have been able to disagree, challenge each other and spend long hours discussing the direction of each of the projects highlighted within these chapters. We have enjoyed exploring new territory – new techniques, novel material and subjects – and as a result there has been no standard methodology that we have necessarily applied from one project to the next. And, throughout our work together, we have respected each other's independence, simply allowing each other to run with ideas and use appropriate skills. Consequently, we have each ensured that the others have had room to move and experiment, while trusting that our goal in each project is completely and passionately shared – though it can, and often did, evolve.

Where one project might focus on the anatomy of a single animal or even the form of a single tooth, another would require detailed skeletal reconstructions of several fossil species. Other projects again have needed the rendering of landscapes with fully-fleshed animals and plants set against a backdrop of environmental conditions estimated for the times – perhaps 110 million years ago in some cases, or 560 million in others. By comparing and contrasting our present with evidence from vast passages of geological time, we have gained extraordinary insights. Seemingly only yesterday, the megafauna that once graced the Australian landscape vanished as a result of simultaneous human occupation and climate change. We have been left to wonder why, and with each attempt to find answers we have progressed a little in our understanding and uncovered further questions to test and refine our knowledge.

What has been so clear to us from the outset is that, just like the chance preservation and sequence of events which have ensnared past life and transformed it into fossils, our projects have often been the result of chance. The odds of finding a fossil might often seem low, but there have been ways of dramatically improving the chances of discovery. So too, more often than not, the chance of each of our projects going ahead has been encouraged and fostered by one or more of us first recognising, and then seizing on, an opportunity that did not just walk into our lives. We chased that chance and made it reality. Three sets of Australia Post stamp issues on Australian palaeontology, for example, began with a single opportunity – the release of Steven Spielberg's *Jurassic Park*. Two further issues were organised through our active pursuit, which provided the funding that resulted in the reconstructions for the Precambrian biota and the megafauna. Without that funding provided by Australia Post, those visual syntheses would not have eventuated. These were the successes. There were also blind alleys, where ideas were floated but, because of the lack of funding and/or lack of time, the texts and the images were never developed.

There has been no standard approach to procuring funding for either the research or the illustration. Each project has been unique – many falling outside the usual mechanisms available via educational and scientific institutional spheres or the commercial and business world where funds could be sought. While each of these sources has provided substantial support, the mix has varied greatly. It has been vital to be able to bring on board the right organisations and people to help us realise our goals. These have included an impressive number of volunteers and art collectors, many of whom have personally provided funding and generously devoted time to make some projects work.

We have learned that structure and breadth of application of projects are factors worthy of consideration if the outcomes are to be financially self-sustaining and even profitable. Each of us has been conscious that, where some projects have been successful, the proceeds, financial or otherwise, can be redirected to drive and support ones that are less so. Some projects clearly have commercial imperatives, whereas others could never have that potential – and nor is it desirable that they should.

Collectively and individually, our value judgements for the worth of a given proposal have never been formulaic or biased heavily towards either the 'commercial' or the 'academic'. It seems that the world is just too interesting in so many different directions. Even if we did not realise that at the beginning of any one of our collaborations, our individual enthusiasms have been thoroughly infectious.

Fortunately, salaries can provide a firm base for scientific and education careers, but funding for the artist is more problematic. Peter has taken a much more risky path, than Pat and Tom, for his income has been closely tied to each ongoing project. He has not sought an institutional or commercial artistic career, choosing instead to freelance. This has necessitated undertaking a wide variety of work, but it has also allowed him the freedom to paint without boundaries or predefined objectives in other pursuits. Despite these differences, we each knew that our collaboration would mean taking risks – professional, personal or financial – through all of which we remained mutually supportive.

There has been an unspoken, and very simple, unifying philosophy, independent of the nature or the dimension of the task – and that is to give of our best. Perhaps we intuitively recognised this resolve in each other from the beginning. We hope that the results of this shared vision are apparent in this book.

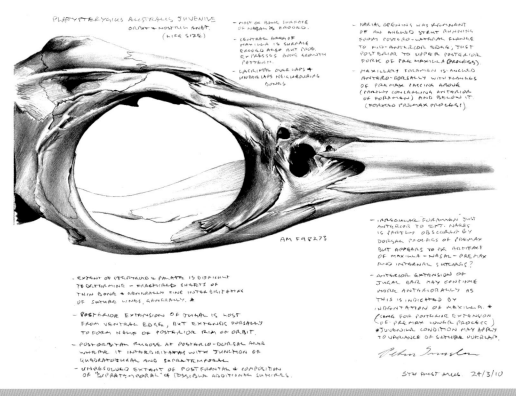

A work in progress. Peter's sketch of the fossil Ichthyosaur in the South Australian Museum, Adelaide. 'Young Ichthyosaur skull' (2010): study notes of *Platypterygius australis* juvenile skull specimin. Graphite on paper, 38 x 28 cm, collection of the artist (P. Trusler).

Glossary

acritarch a group of decay-resistant, organic-walled microorganisms. Most from the Proterozoic and Paleozoic are interpreted as single-celled, photosynthetic organisms. Some researchers suggest they are related to a living group called dinoflagellates.

ambush predator takes live prey by lying in concealment until the prey happens to pass by. This is distinct from 'pursuit predators', which actively hunt for and run down their chosen prey.

ammonoid extinct, shelled cephalopods with septa that are convex in the direction of the opening rather than concave as in nautiloids. Geological range: Early Devonian to Late Cretaceous.

anagalid extinct mammals of uncertain affinities but considered to possibly be related to rodents and/or rabbits and/or elephant shrews.

anhingas a group of water birds related to cormorants. They possess a long neck, snake-like in appearance, and are also known as darters or snake-birds.

anoxic 'without oxygen'. Usually applied to ecological conditions where dissolved oxygen has been depleted from water.

anseriform waterfowl; comprising ducks, swans, geese and screamers.

anthropomorphic possessing human form or characteristics.

anthropomorphism: attributing a phenomenon, objects or animals with human motivation or character.

arboreal lives in trees.

arthrodire 'jointed neck'; a prominent member of armoured fish call placoderms that dominated fish faunas of the Devonian period. Characterised by two pairs of upper tooth plates rather than one.

arthropod any one of a group of solitary, marine, freshwater or aerial invertebrates belonging to the order Arthropoda, which are characterised chiefly by joined appendages and segmented bodies.

artiodactyl a highly diverse order of mammals, most typically having two or four hooves on each foot. Sheep, goats, cattle, deer, antelope and pigs.

avifauna bird species found in a particular area or throughout a given time period.

basicranium base of skull.

benthic pertaining to the bottom of the ocean or lake. May include the sediment layers and the water immediately above it as an ecological zone.

biomat a living layer of microbes (mainly of bacteria or algae) which can cover a surface – in Precambrian times biomats covered expanses of the ocean floor.

biota the life found in a particular area or characteristic of a particular period of time.

brachiopod marine invertebrates enclosed in two shells, but not related to bivalve molluscs. Although bilaterally symmetrical, the lower shell is not identical to the upper shell and may contain a hole for a short attachment stalk. Early Cambrian to the Present.

brachycephalic 'short head'; typically exemplified by a broad face.

branchial basket a basket-like internal structure that supports the gills in lower chordates. In tunicates it also acts as a filter to entrap plankton food from water.

bryozoan 'moss animals'; colonial aquatic organisms that superficially resemble corals. However, many forms do not have mineralised skeletons and they are structurally more complicated in having three tissue layers rather than just two. Ordovician to Recent.

camera lucida a simple optical device allowing the user to view an image with one eye and a drawing surface with the other, and thereby trace the image directly onto the surface. Especially useful in microscopy for directly 'tracing' the view of a magnified object.

cathartid the 'New World' vultures of North and South America.

cephalopod 'head foot'; the only molluscs with a definite head. Includes squid, cuttlefish, octopus, nautilus, ammonites and belemnites. Cambrian to Recent.

clast a particle such as a sand grain or pebble in a sedimentary rock.

climax (ecology) the final stable stage of development of a community, species, flora or fauna in a given environment

cnidarian corals and their relatives; sea anemones, sea-pens, jellyfish and hydra.

cursorial to run.

dimorphism, sexual the phenomenon where the two genders of a species have a different appearance.

dinophile 'terrible lover'; a person who is an ardent student of dinosaurs.

dorsal the top of an object.

dromornithid belonging to a family of extinct, flightless, herbivorous birds related to waterfowl. Confined to Australia, they flourished from 50 million years ago until as recently as 50 000 years ago.

ecocline a gradient in community structure that correlates to one or more physical changes in the environment over a geographical area.

edentate 'without teeth'; applied to a group of mammals, the Edentata, which includes tree sloths, anteaters, armadillos, extinct ground sloths and glyptodonts. While some are toothless, most possess teeth. However, those teeth often lack enamel.

ediacarans multicellular organisms that lived during the Ediacaran geological period, 630–542 million years ago.

enantiornithine primitive birds that lived during the Mesozoic Era.

endocast cast of the brain case of a vertebrate, or an internal cast of some part of an organism.

endothermy being warm blooded.
estuarine the environment where the fresh water of a river mixes with the salt water of the ocean.
eupantothere primitive group of mammals out of which the marsupials and placentals arose during the Mesozoic Era.
fenestra/e small hole/s in bone allowing for the passage of nerves and blood vessels.
filiform shaped like a thread or filament.
floristics the study of the distribution of plants over an area.
foramen/foramina: natural hole/s in bone allowing for the passage of nerves and blood vessels.
fossa depression or natural excavation in a bone surface.
frondose bearing fronds or shaped like a finely divided leaf, as in ferns and palms.
frontal bone at the front and top of the skull of a vertebrate.
fusiform spindle or fish-shaped. Wide in the middle and tapering to each end.
gastrolith rock ingested to facilitate the breakdown of food in the digestive tract.
gouache opaque water-based paint, often known as designer's or poster paint.
gruiform crane-like; a major group of birds that include cranes, rails and their relatives.
gymnosperm a plant such as a conifer or cycad, the seeds of which are not formed within a fruit.
gypaetine belonging to the Old World vultures, such as the griffons.
hadrosaur 'duckbilled' dinosaur; bipedal, herbivorous dinosaurs that thrived during the Cretaceous Period. Occurred on every continent except Australia.
homology the recognition of structures in different organisms that have a common origin in a form ancestral to both.
ikatite or ikaite a mineral formed in seawater close to its freezing point.
integument/ary the external surface of an organism, the skin or cuticle and its out-growth structures.
labyrinthodont 'labyrinthine' teeth; a superseded name for temnospondyl amphibians that possessed teeth with convoluted, folded enamel.
lagerstätte (pl. lagerstätten) a fossil locality which is highly remarkable for either its diversity or quality of preservation; sometimes both.
lamina (pl. laminae) Latin for 'plate'; used in reference to structure having the form of thin sheets.
lithify/lithification 'to become rock'; the process of becoming rock, as when loose sand becomes hardened into sandstone by either compaction and/or the cementing of the individual grains together by the deposition of a natural cement such as calcium carbonate.

lithograph literally, 'writing on stone'; a term applied to illustrations on paper printed from images drawn on limestone.
megascopic an item or organism large enough to be seen with the naked eye.
metazoan an animal composed of more than one cell.
mihirung Australian Aboriginal word that seems to have referred to the Dromornithidae, large, extinct, flightless birds confined to Australia. Now used as venacular name for the Dromornithidae.
mitochondrial (genes)/mitochondria 'bodies' within a cell where the metabolic functions are carried out; genes – chemical sequences on DNA molecules within chromosomes and also on RNA molecules that regulate the functions of the cell in which they are located.
monophyletic said of a group of organisms that share a common ancestor.
monotreme egg-laying mammals, the living representatives being the platypus and echidna.
morphology the study of the shape or form of organisms.
multituberculate an extinct order of mammals that resembled some living rodents but are not particularly closely related to them.
nautiloid shelled cephalopods with internal partitions or septa that are concave in the direction of the opening, rather than convex as in ammonoids. *Nautilus* is the only living cephalopod with a shell. Geological range: Cambrian – Recent.
neognathous the mobile palatal structure of the skull of all living birds other than the ground-dwelling ratites, such as the emu and ostrich.
occlusal view to look upon the crown of a tooth as a piece of food about to be consumed would view it – if it could.
ontogenetic the development that an organism goes through from embryo to adult.
ornithomimosaur bipedal, carnivorous dinosaurs, most of which lack teeth.
ornithopod bipedal, herbivorous dinosaurs.
orthographic projection a method of drawing in which all the lines projected from the object to the observer are parallel and perpendicular to the plane of view. This is distinct from a perspective view or convergent projection. The orthographic method provides no sense of depth but allows accurate measurements to be taken.
osteology the study of bones.
oxbow small lake located in a former meander loop cut off from the main stream of a river. Eventually, oxbow lakes silt up to form marshes and finally meander scars. Some but not all billabongs are oxbows.
palaelodid an extinct group of waterbirds that existed during the Cenozoic.

palaeognathous the rigid skull structure characteristic of the ground-dwelling ratite birds such as the emu, kiwi and ostrich.

palaeoniscid the most primitive group of bony fish.

palynology the study of pollen and spores, especially their fossils.

parallax the change in appearance and/or apparent position of an object by viewing it from different lines of sight. The phenomenon can be used to calculate distances and is the means by which depth perception is attained through stereoscopic sight.

parietal a bone in the top rear region of a skull.

pectinate comb-like or possessing a row of tooth-like projections or prongs.

pennatulacean sea pens, a relative of corals.

petrification 'turning to stone'; the processes by which the remains of a once living organism are transformed into a fossil. Commonly the replacement of organic material by minerals like silica.

petrosal the bone in the skull that contains the sensory tissue for both hearing and maintaining balance.

phylogeny the study of the pattern of relationships between organisms; a study of 'family trees'.

pika short-eared rabbits; also known as rock rabbits or coneys.

placental mammal mammals whose young are born at an advanced stage of development, in contrast to marsupials, which are born at a much more immature stage.

placoderm 'armoured skin'; a group of fish that originated in the Silurian, thrived in the Devonian and became extinct in the Early Carboniferous. Characterised by armour on the head and trunk, paired fins, and a hinged jaw.

platanistid 'river dolphins'; belonging to the family that includes the Ganges River dolphin.

postcranial all bones of a skeleton other than the skull.

precocial said of animals that when born are independent, as opposed to having atrical young that require parental feeding, care and protection.

proprioceptive the sense by which we perceive the relative position or state of our body parts and organs.

rangeomorph a very successful group of organisms that existed during the Neoproterozic which had a repeating structure, often regarded as fractal.

ratites large, flightless birds such as the ostrich, emu and kiwi.

reptile radiation the family tree of reptiles or the process of reptile diversification.

reticulate a network, or resembling a network.

ribosomal RNA ribosomal ribonucleic acid (rRNA) is the central component of the ribosome, the protein manufacturing machinery of all living cells. Its function is to provide a mechanism for decoding mitochondrial RNA (mRNA) into amino acids.

rostrum in anatomy, a beak-like or forward extension of the face, nose and/or jaws.

sagittal the vertical plane parallel with the longitudinal axis of a skull that divides left from right.

saurian reptile.

sauropod herbivorous, quadrupedal dinosaurs characterised typically by a small skull, long neck and tail; includes the largest terrestrial vertebrates that ever lived.

sclerophyllous bearing hard leaves; plants with hard, waxy, leaf surfaces typical of arid and semi-arid environments.

scutes plates embedded in the skin of an animal.

sessile said of an organism that does not move.

stromatolite a laminar rock formed by the precipitation of carbonates or silicates or the entrapment of sediment by microorganisms, commonly cyanobacteria.

taphonomy the study of the processes by which the remains of an organism become a fossil.

taxa natural groups of organisms.

taxonomy the classification of organisms.

temnospondyls early tetrapods which may have given rise to the living amphibians. They were a very diverse group that flourished from the Early Carboniferous to the Early Cretaceous. Many of the Palaeozoic temnospondyls were terrestrial, however in the Mesozoic most were aquatic.

thecodont teeth which are anchored to the jaw by means of roots inserted into sockets.

thermoregulation maintenance of body temperature.

theropod 'beast feet'; bipedal 'lizard-hipped' dinosaurs. The group was mostly carnivorous, and famously includes *Tyrannosaurus rex*. Small specialised forms gave rise to birds during the Jurassic.

trilobite extinct marine arthropods characterised by three longitudinal lobes and three transverse ones.

tunicate a group of animals which have a mobile larva, but as adults sit attached to the seafloor. Also called sea squirts, for if disturbed they literally contract and squirt water. Despite appearing primitive, they belong to the urochordates and are related to vertebrates.

varanids carnivorous lizards.

vascularise/d to have tubes/the presence of tubes; typically meaning to contain blood vessels.

vibrissae sensory hairs on the snout.

References

Chapter 2: The Dragon Chasers
Rich, P. V., 1973. The History of Australia's Non-Passeriform Birds. PhD thesis, Columbia University, New York (published as several papers).
Rich, T. H., 1972. Deltatheridia, Carnivora, and Condylarthra (Mammalia) of the Early Eocene, Paris Basin, France. *University of California Publications in the Geological Sciences* 88:1–72 (Masters thesis publication).
Rich, T. H., 1981. Origin and history of the Erinaceinae and Brachyericinae (Mammalia, Insectivora) in North America. *Bulletin of the American Museum of Natural History* 171:1–116 (PhD thesis publication).

Chapter 3: The Beginning – Bacchus Marsh *Diprotodon*
Gary, M., McAfee, R and Wolf, C. L. ,1972. *Glossary of Geology.* American Geological Institute, Washington, D. C.
Kloot, T. and McCulloch, E. M., 1980. *Birds of Australian Gardens.* Rigby, Melbourne.
Owen, R., 1838. Fossil remains from Wellington Valley, Australia. In Mitchell, T. L., *Three Expeditions into the Interior of Eastern Australia, with Descriptions of the Recently Explored Region of Australia Felix, and the Present Colony of New South Wales.* Vol 2. T. & W. Boone, London.
Owen, R., 1870. On the Fossil Mammals of Australia – Part III. *Diprotodon australis*, Owen. *Philosophical Transactions of the Royal Society of London*, 160: 519–78.
Owen, R., 1877. *Researches on the Fossil Remains of the Extinct Mammals of Australia with a Notice of the extinct marsupials of England.* J. Erxleben, London.
Rayfield, E. J., 2004. Cranial mechanics and feeding in *Tyrannosaurus rex*', *Proceedings of the Royal Society B. Biological Sciences, 271*, (pp. 1451–9).
Rayfield, E. J., 2007. Finite element analysis and understanding the biomechanics and evolution of living and fossil organisms. *Annual Review of Earth and Planetary Science Letters* 35:541–76.

Chapter 4: The Art of *The Fossil Book*
Fenton, C. L. and Fenton, M. A., 1958. *The Fossil Book. A Record of Prehistoric Life.* Doubleday, Garden City.
Rich, P. V., Rich, T. H., Fenton, M. A. and Fenton, C. L., 1989. *The Fossil Book. A Record of Prehistoric Life.* Doubleday, New York.
Rich, P. V., Rich, T. H., Fenton, M. A. and Fenton, C. L., 1996. *The Fossil Book. A Record of Prehistoric Life.* Dover Publications, Mineola.
Rich, P. V., Rich, T. H., Fenton, M. A., Fenton, C. L., Bondarenko, O. B., Golubev, V. N., Gubin, Y. M., Esin, D. N., Koznetsova, T. V., Kurotchkin, E. N., Michailova, I. A., Naugolnich, S. B. and Rozanov, A. Y., 1997. *Kamennaya Kniga* (*The Fossil Book*), Mezhdunarodnaya Akademicheskaya Izdateliskaya kompaniya "Nauka", Moscow (*The Fossil Book*, Russian Edition).
Rich, P. V., van Tets, G. F. and Knight, F., 1985. *Kadimakara. Extinct Vertebrates of Australia.* Pioneer Design Studio, Lilydale.
Rich, P. V., van Tets, G. F. and Knight, F., 1990. *Kadimakara. Extinct Vertebrates of Australia.* Princeton University Press, Princeton.
van Tets, G. F., Rich, P. V. and Marino-Hadiwardoyo, R., 1989. A reappraisal of *Protoplotus beauforti* from the Early Tertiary of Sumatra and the basis of a new pelecaniform family. *Publication of the Geological Research and Development Centre, Paleontology Series, Ministry of Mines and Energy, Geological Research and Development Centre*, 5: 57–75.

Chapter 5: Dinosaurs from China
Hopson, J. A., 1975. The evolution of cranial display structures in hadrosaurian dinosaurs. *Paleobiology*, 1: 21–43.
Komarower, P., 2002. *The Development of Vertebrate Palaeontology in China During the First Half of the Twentieth Century.* PhD thesis, Monash University, Melbourne, Australia.
McClellan, E. et al., 1982. *Dinosaurs from China.* Council of the National Museum of Victoria, Melbourne.
Ouyang, H. and Ye, Y., 2002. *The First Mamenchisaurian Skeleton with Complete Skull: Mamenchisaurus youngi.* Sichuan Science and Technology Press, Chengdu.
Rich, P. V., Zhang, Y. P., Chow, M. C., Wang, B. Y., Komarower, P., Fan, J. H., Sloss, R., Moody, J. K. M. and Dawson, J., 1994. *A Chinese–English and English–Chinese Dictionary of Vertebrate Palaeontology Terms.* Monash University, Melbourne.
Vickers-Rich, P., 1996. Yuping: Palaeontology in China. *Ormond Papers* 1996: 149–54.
Weishampel, D. B. and Horner, J. R., 1990. Hadrosauridae. In Weishampel, D. B., Dodson, P. & Osmolska, H., eds. *The Dinosauria.* University of California Press, Berkeley: 534–61.
Young, C.-C., 1958. The dinosaurian remains of Laiyang, Shantung. *Paleontologica Sinica*, Series C, 16: 1–138 (in Chinese and English).
Young, C.-C. and Chao, H.-C., 1972. *Mamenchisaurus hochuanensis* sp nov. *Institute of Vertebrate Paleontology and Paleoanthropology Monographs*, Series A, 8: 1–30 (in Chinese).

Chapter 6: *Wildlife of Gondwana*

Campbell, K. S. W. and Barwick, R. E., 1987. Palaeozoic lungfishes: A review. *Journal of Morphology, Supplement*, 1: 93–132.
Hecht, M., 1975. The morphology and relationships of the largest known terrestrial lizard, *Megalania prisca* Owen, from the Pleistocene of Australia. *Proceedings of the Royal Society of Victoria* 87: 239–50.
Jenkins, F. A. and Goslow, G. E., 1983. The functional anatomy of the shoulder of the Savannah Monitor Lizard (*Varanus exanthematicus*). *Journal of Morphology*, 175:195–216.
Long, J. A., 1988. Late Devonian fishes from Gogo, Western Australia. *National Geographic Research*, 4 (4): 436–50.
Molnar, R. E., 2004. *Dragons in the Dust. The Paleobiology of the Giant Monitor Lizard* Megalania. Indiana University Press, Bloomington.
Rich, T. H. and Hall, B., 1979. Rebuilding a giant. *Australian Natural History*, 19 (9): 310–13.
Rich, T. H. and Rich, P. V., 1989. Polar dinosaurs and biotas of the Early Cretaceous of southeastern Australia. *National Geographic Research*, 5 (1): 15–53 (the naming of *Leaellynasaura*).
Stirling, E. C. and Zietz, A. H. C., 1896. Preliminary notes on *Genyornis newtoni*: A new genus and species of fossil struthious bird found at Lake Callabonna, South Australia. *Transactions and Proceedings of the Royal Society of South Australia*, 20: 171–90.
Vickers-Rich, P. and Rich, T. H., 1993. *Wildlife of Gondwana. The 500-Million-Year History of Vertebrate Animals from the Ancient Southern Supercontinent*. Reed Books, Sydney.
Vickers-Rich, P. and Rich, T. H., 1999. *Wildlife of Gondwana. Dinosaurs & Other Vertebrates from the Ancient Supercontinent*. Indiana University Press, Baltimore.
Webb, G. J. W. and Gans, C., 1982. Galloping in *Crocodylus johnsoni* – a reflection of terrestrial activity. *Records of the Australian Museum*, 34 (14): 607–18.
White. M. E., 1986. *The Greening of Gondwana*. Reed, Sydney.

Chapter 7: The Dinosaurs of Darkness

Chinsamy, A., Rich, T. H. and Vickers-Rich, P., 1998. Polar dinosaur bone histology. *Journal of Vertebrate Palaeontology* 18:385–90.
Douglas, J. G. and Williams, G. E., 1982. Southern polar forests: The Early Cretaceous floras of Victoria and their palaeoclimatic significance. *Palaeogeography, Palaeoclimatology, Palaeoecology*, 39: 171–85.
Haines, T., 1999. *Walking With Dinosaurs. A Natural History*. BBC Worldwide, London (book and BBC Television series. Episode/Chapter 5 Spirits of the Ice Forest was about *Leaellynasaura* and her world during the Cretaceous).
McNamara, G. C. and Vickers-Rich, P., 2002. *Dinosaurs of Southern Australia*. Geoscience Australia, Record 2002/09 (a teaching kit with images for use in primary and secondary schools).
Rich, L., 1994. My little dino. *Ranger Rick*, 29: 32–9.
Rich, T. H., 2007. *Polar Dinosaurs of Australia*. Museum Victoria, Melbourne.
Rich, T. H. and Rich, P. V., 1989. Polar dinosaurs and biotas of the early Cretaceous of southeastern Australia. *National Geographic Society Research Report* 5: 15–53.
Rich, P. V., Rich, T. H., Wagstaff, B., McEwen Mason, J., Douthitt, R. T., Gregory, R. T. and Felton, A., 1988. Evidence for low temperatures and biologic diversity in Cretaceous high latitudes of Australia. *Science*, 242: 1403–6.
Rich, T. H. and Vickers-Rich, P., 2000. *Dinosaurs of Darkness*. Indiana University Press, Bloomington.
Rich, T. H. and Vickers-Rich, P., 2001. *Dinosaurs of Darkness*. Allen & Unwin, Sydney.
Rich, T. H. and Vickers-Rich, P., 2003. Protoceratopsian? ulnae from Australia. *Records of the Queen Victoria Museum*, Launceston, 113: 1–12 (*Serendipaceratops arthurcclarkei* – named after Arthur C. Clarke).
Warren, A., Rich, T. H. and Vickers-Rich, P., 1997. The last last labyrinthodont? *Palaeontographica Abteilung*, A, 47: 1–24.

Chapter 8: A Moa Mummy – A classic dissection

Bock, W. J., 1963. The cranial evidence for ratite affinities. In: Sibley, C. G., ed. *Proceedings XIII International Ornithological Congress*, Ithaca, 17–24 June 1962, 1: 39–54.
Cooper, A., et al., 1992. Independent origins of New Zealand moas and kiwis. *Proceedings of the National Academy of Sciences*, 89 (18): 8741–4.
Murray, P. & Vickers-Rich, P., 2004. *Magnificent Mihirungs. The Colossal Flightless Birds of the Australian Dreamtime*. Indiana University Press, Bloomington.
Vickers-Rich, P. & Rich, T. H., 1999. *Wildlife of Gondwana. Dinosaurs & Other Vertebrates from the Ancient Supercontinent*. Indiana University Press, Bloomington.
Vickers-Rich, P., Trusler, P., Rowley, M. J. H., Cooper, A., Chambers, G. K., Bock, W. J., Millener, P., Worthy, T. and Yaldwin, J. C., 1995. Morphology, myology, and DNA of a mummified Upland Moa, *Megalapteryx didinus* (Aves: Dinornithiformes) from New Zealand. *Tuhinga, Records of the Museum of New Zealand Te Papa Tongarewa*, 4: 1–26.
Worthy, T. H., 1988. A re-examination of the moa genus *Megalapteryx*. *Notornis*, 35 (2): 99–108.

Chapter 9: Magnificent Mihirungs

Gregory, J. W., 1906. *The Dead Heart of Australia. A Journey Around Lake Eyre in the Summer of 1901-1902, With Some Account of the Lake Eyre Basin and the Flowing Wells of Central Australia*. John Murray, London.
Mayr, E. and Stein, G. H. W., 1944. The birds of Timor and Sumba. *Bulletin, American Museum of Natural History*, 83: 127–94.
Murray, P. & Vickers-Rich, P., 2004. *Magnificent Mihirungs. The Colossal Flightless Birds of the Australian Dreamtime*. Indiana University Press, Bloomington.
Rich, P. V., 1979. The Dromornithidae, an extinct family of large ground birds endemic to Australia. *Bulletin of the Bureau of Mineral Resources, Geology and Geophysics*, 184: 1–188.
Rich, P. V., 1980. The Australian Dromornithidae: A group of extinct large ratites. *Contributions to Science, Natural History Museum of Los Angeles County*, 330: 93–103.
du Toit, A. L., 1937. *Our Wandering Continents: A Hypothesis of Continental Drifting*. Oliver & Boyd, London.
Wegener, A. L., 1915. *Die Entstehung der Kontinente und Ozeane* Braunschweiz, Vieweg.
Wegener, A. L., 1924 *The Origin of Continents and Oceans*. Methuen, London. (First English translation.)

Chapter 10: Where the Wind Bites – Patagonia

Rauhut, O. W. M., Caldera, G., Vickers-Rich, P. and Rich, T., 2003. The oldest Cretaceous dinosaurs from Chubut Province, Argentina. *Cretaceous Research*, 24: 487–97.
Rich, T. H., Vickers-Rich, P., Gimenez, O., Cuneo, R., Puerta, P. and Vacca, R., 1999. A new sauropod dinosaur from Chubut Province, Argentina. *Proceedings 2nd Gondwanan Dinosaur Symposium, National Science Museum Monograph*, 15, Tokyo: 61–-84.

Chapter 11: *The Rise of Animals* – Back to the Precambrian

Fedonkin, M. A., Gehling, J., Grey, K., Narbonne, G. and Vickers-Rich, P., 2007. *The Rise of Animals. Evolution and Diversification of the Kingdom Animalia*. Johns Hopkins University Press, Baltimore.
Sano, S., Terada, K., Vickers-Rich, P. and Trusler, P., 2006. *Before the Dinosaurs. The First Animals on Earth*. Fukui Prefectural Museum Exhibition Catalogue and Exhibition, Fukui (in Japanese).
Schopf, W. and Klein, C., 1992. *The Precambrian Biosphere: A Multidisciplinary Study*. Cambridge University Press, Cambridge.
Vickers-Rich, P. and Komarower, P., 2007. *The Rise and Fall of the Ediacaran Biota*. Geological Society of London, Special Publication 286.
www.geosci.monash.edu.au/precsite (website for International Geological Correlation Project 493, The Rise and Fall of the Ediacaran (Vendian) Biota, UNESCO).

Chapter 12: Chinese and Australian Mesozoic Mammals

Chow, M. & Rich, T. H. V., 1982. *Shuotherium dongi*, gen. et sp. nov., a therian with *pseudotribosphenic* molars from the Jurassic of Sichuan, China. *Australian Mammalogy*, 5: 127–42.
Crompton, A. W., 1971. The origin of the tribosphenic molar. In Kermack, D. M. and Kermack, K. A., eds. Early mammals. *Linnean Society, Zoology*,. 50: Supplement 1: 65–87.
Rich, T. H., 2008. The Palaeobiogeography of Mesozoic mammals: A review. *Arquivos do Museu Nacional*, Rio de Janeiro, 66: 231–49.
Rich, T. H., Hopson, J., Musser, A., Flannery, T. and Vickers-Rich, P., 2005. Independent origins of the middle ear bones in monotremes and therians. *Science*, 307: 910–14.
Rich, T. H. & Vickers-Rich, P., 2000. *The Dinosaurs of Darkness*. University of Indiana Press, Bloomington.
Rich, T. H., Vickers-Rich, P., Constantine, A., Flannery, T. F., Kool, L. and van Klaveren, N., 1997. A tribosphenic mammal from the Mesozoic of Australia. *Science*, 278: 1438–42.
Rich, T. H., Vickers-Rich, P., Flannery, T. F., Kear, B. P., Cantrill, D., Komarower, P., Kool, L., Pickering, D., Trusler, P., Morton, S., van Klaveren, N. and Fitzgerald, E., 2009. A multituberculate from the Cretaceous of Australia. *Acta Palaeontologica Polonica*, 54: 1–5.
Wang, Y, Clemens, W. A., Hu, Y. et al., 1998. A probably *pseudo-tribosphenic* upper molar from the Late Jurassic of China and the early radiation of the Holotheria. *Journal of Vertebrate Paleontology*, 18: 777–87.

Chapter 13: The Last of the Mob

Archer, M., 1982. *Carnivorous Marsupials*. Surrey Beatty & Sons and the Royal Zoological Society of New South Wales, Sydney.
Clark, R. and Colligan, M., 1993. *The Collection of 1993. Australian Stamps*. Australia Post, Melbourne.
Fedonkin, M. A., Gehling, J., Grey, K., Narbonne, G. M. and Vickers-Rich, P., 2007. *The Rise of Animals. Evolution and Diversification of the Kingdom Animalia*. Johns Hopkins University Press, Baltimore.
Flannery, T. F., 1994. *The Future Eaters: An Ecological History of the Australasian Lands and Peoples*. Reed Books, Sydney.
Kloot, T. and McCulloch, E., 1980. *Birds of Australian Gardens*. Rigby, Melbourne.

Murray, P. and Vickers-Rich, P., 2004. *Magnificent Mihirungs. The Colossal Flightless Birds of the Australian Dreamtime.* Indiana University Press, Baltimore.
Simpson, K., Day, K. and Trusler, P., 2004. *Field Guide to the Birds of Australia.* 7th edition. Christopher Helm, Melbourne.
Vickers-Rich, P., 2008. *From Small Things Big Things Grow.* Australian Postal Corporation, Melbourne.
Vickers-Rich, P., Monaghan, J. M., Baird, R. F. and Rich, T. H., 1996. *Vertebrate Palaeontology of Australasia.* Monash University Publications Committee, Melbourne.
Vickers-Rich, P. and Trusler, P., 2008. *Megafauna in Australia.* Australian Postal Corporation, Melbourne.
Wells, R. T. and Nichol, B., 1977. On the manus and pes of *Thylacoleo carnifex* Owen (Marsupialia). *Transactions of the Royal Society of South Australia,* 101 (5-6): 139–46.

Chapter 14: The Art of Humour and Enchantment

Ramos-Horta, J. and Vickers-Rich, P., 2009. The Lost World of Timor-Leste. A Boy and a Crocodile Travel Through Time. Monash University, Monash Science Centre, Melbourne (two books, one simple, one complex, in Tetun, Portuguese, English, Chinese and Spanish on the geology of Timor for children).
Vickers-Rich, P., Rich, L. S. V. and Rich, T. H., 1996. *Australia's Lost World. A History of Australia's Backboned Animals.* Kangaroo Press, Sydney.

Index

Page numbers in bold refer to illustrations

Aboriginal culture, 69
acid etching (preparation), **97**
Africa, 148, 265
'Agronomic Revolution', **231**
Alcoota, 163, 166, 167–9, 174
Alerces National Park, **189, 190, 191, 192**
All about Dinosaurs, 14
Allocasuarina, 96
Allosaurus, 120
Alola Foundation, 285
American Museum of Natural History (AMNH), 16–17, 27, 164–5
ammonites, 46
anagalids, **56**
anhigas, 52
    See also snake birds
Anhimidae, 169
Anseranas, 169
anseriforms, 169
Antarctic Circle, 108
Antarctica, 17, 54, **55**, 120, 122, 184, 265
ANZAAS (Australia New Zealand Association for the Advancement of Science), 78, 84
Araucaria araucana, **192**
araucarian, **187**, 188
Archangel'sk, 204, 207
Archbold, N., 99
Archer, M., 109–10
Arctic, **209**
Argentina, 82, **183, 189, 190, 191, 192,** 193–4, **196, 197**
Arminiheringia, **82**
artistic representations
    coloration, 92
    cultural distortion, 128
    cultural factors, 132
    emotional element, 72–3
    graphic skills, 39
    motion, 74
Asa Zoo, 106, 138
Asia, 265
'Asian Cainozoic fauna', **56**
Atlascopcosaurus, **124–5**
Attenborough, D., 211
Audubon, J.J., 37
'Aurora', **104**, 122–3, 251
Ausia, 207, 209
ausktribosphenids, **249**, 251, 290
Ausktribosphenos, 248
'*Ausktribosphenos nyktos* – mandible', **233**
Australasian avifauna, 164
Australia, 120, **124–5**, 169, 184, 186, 201, 213–14, 216, 220, 225, 230, **252–3**, 257, 289

monsoon country, 169
Australia Post, 67, 83, 193, 200, 206, 214, 224–5, 256–7, 295
    Australian Megafauna philatelic issue, **258–9, 263, 268, 271, 272, 273, 277, 279, 281**
    Australia's Dinosaur Era philatelic issue, **118–19, 124–5**, 264
    Creatures of the Slime philatelic issue, **198**
    Philatelic Group, 256–7, 260–3, 265
'Australia's Dinosaur Era', **124–5**
Australian Aboriginal mythology, **165**
Australian Academy of Sciences, 47
Australian Dinosaurs stamps, *see* Australia Post
Australian Emperor Dragonfly, **141**
'Australian megafauna', 67, **258–9**
Australian Museum (Sydney), 66, 82, 98
Australian Natural History, 88
Australian Research Council, 20
Australian Research Grants Committee, 24, 26–7
Australosphenida, 242
Avalon Peninsula, Newfoundland, **215, 223**, 230
avian relationships, 168

Bacchus Marsh, 24–5, **26**, 29, 30, 33, 68, 261
Ballarat Field Naturalists Club, 8
Bartholomai, A., 17–18
Benitez, A, 186
'*Beothukus* – Newfoundland', **226**
Bierstadt, A., 129
billabong, 111, **112–13**
binocular depth perception, 43
bio-illustration, 40
biomat, 216, **230**
Birds of Australian Gardens, 40, 261
Birds of Prey, 224
'Bishops', **252–3**
'Bishops – hypothetical posture studies', **249**
Bishops whitmorei, **239, 241**, 242, 248, **251, 252–3**
Bock, W., 149, 164–5
borhyaenid marsupial, **82**
brachiopods, 286
'*Bradgatia* – partial reconstruction, Newfoundland', **224**
Brodkorb, P., 16
Brown, L. (Capability), 129
browsing kangaroo, **59**
bryozoans, 286
Buenos Aires, 193
Bullock Creek, 148, 168–9, 174
Bullockornis planei, 166

Bushman habitations, 207, **212**
Bushman rock art, **212**
Bushy-tail Blowout (Wyoming), 193

calcite, 108
Callitris, 96
Cambrian, 219
Camens, A., 262, 274
Campbell, K., 98–9
Capello, M., 184, 188, 193
Carey, S., 274
cassowaries, 92, 148
cathartid vultures, 163
cave art, 265
Central Australia, 54, **59**, 84
cephalopods, 99, 195
Ceratodus, **143**, 144
'Cerro los Chivos', **196**
Charnia, 219
Charniodiscus, **218**, 219
Chasmatosaurus, **55**
Childs Frick Laboratory, 16, 165
China, 46–7, 54, 64, 65, 67–9, 144, 186, 203, 234, 240
    Cultural Revolution, 46
Chinese–English Dictionary and English–Chinese Dictionary of Paleontology terms, **47**, 47–8, 65
Chinsamy, A., 123
Chubut Province (Patagonia), 182–5, **183**, 193, **196, 197**
Chubut River, 184, 188, **197**
Cifelli, R., 242
Cimolesta, **56**
Clark, R., 256
Clarke, A.C., 203
Cleeland, C., 80, 88, 96, 106, 117
cnidarians, 219
Coccolepis woodwardi, **142**
Coccosteus cuspidatus, **100**
Coffa, F., 52, 200, 203
collagen, 149, 154
Columbia University, 16–17, 19, 46, 163–5
condors, **189**
ConocoPhillips, 257, 286
Consejo Nacional de Investigaciones Científicas y Técnicas (CONICET), 193
Constantine, A., 188, 204
continental drift, 164
Cooper, R., 149–50
cormorant, 51–2, **53, 60**
Corriebaatar marywaltersae, 243, **246**
Couzens, J., 66

cranes, 148
Crapp, A., 262
'Creatures of the slime', **198**, 206, 257, 265
Cretaceous mammals, 67
crevasse splay deposit, 111
crinoids, 286
crocodiles, 80, 285
Crompton, A.W., 236
cryptic patterns, 91
CSIRO (Commonwealth Science and Industrial Research Organisation), 51
cultural interaction, 68–9
Cuneo, R., 182–3, 188, 193

Dabis Formation, **228**
Darragh, T., 24–6
darters, 52
Darvall, P., 201–2
Darwin, C., 182
Darwin (NT), 172
Davies, T., 67
Dickinsonia, 290
Dili, 285, 288
Dinornis, 154
dinosaur art, 4, 129
Dinosaur Cove, **12**, 109–11, 116–17, 126, 184, **283**, 290
dinosaurs, 74, 182, 188, 200, 256–7, 264, 270
  animation of, 74
'Dinosaurs from China', 21, 48, **62**, 65, 67, **70**, **71**, **75**, 109
Dinosaurs of Darkness, **104**, 122, **289**, 290
Diprotodon optatum, **2**, **22**, **24**, 25–7, **27**, **28**, 29–30, **31**, **33**, **36**, **39**, **41**, **42**, **43**, 250, **258–9**, 261–2, **275**, **276**, **279**, **280**, 294
diprotodontid track-way, 261, **276**, 280
draftsmanship, 39
Dromornis stirtoni, **160**, 166–7, 169, 172, 174–7, **173–5**, **177**
dromornithids, 16, 18, **79**, 84, 148, **165**, **167**, 169
  footprints, **79**, 261
  palate, 148
drop soil structure, **109**
Droser, M., 214, 220
DSDP (Deep Sea Drilling Project), 17
duck-billed dinosaurs, 116
Dürer, A., 37
du Toit, 50, 164

Eagle's Nest (Vic.), 106, **107**
'Early Miocene lake fauna of central Australia', **60**
Eastmanosteus calliaspis, 98, 99, **102–3**
echidna, 117, 235
ecocline, 172
'Ediacara', **221**
Ediacarans, **199**, 205, **206**, **207**, **208**, 216, **217**, 219, 224–5, 230, 257, 290
  frond-like organisms, 219, 225
education in science, 9

Ektopodon, **59**
elephant birds, 148
Emeriza impetuani, **228**
emus, 148, 151, 169
enantiornithines, 144, **145**
Eocene, 54
'Eocene Messel fauna, Germany', **57**
Eomanis, **57**
Ernanodon, **56**
Erxleben, J., 38
eupantotheres, **244**
Eurotamandua, **57**
Evans, G., 83, 200
Ewing, M., 17

Fahey, S., 263
Fan, J., 67
feathers, 92, 144, **155**, 156
Fedonkin, M., 21, 200, **201**, 202–3, **204**, 205, **206**, 207, 209, 211, 213, 230
Fenton, C.L., **48**, 49, 54
Fenton, M., 46, **48**, 49, 54, 56
Ferguson, W.H., 107, 117
Fernandez Estancia (Chubut Province), 184–5
Field Guide to the Birds of Australia, 261
Filmer-Sanky, P., 153
finite element analysis, 30, **31**
Fitzroya cupressoides, **189**, **190**, **191**
flamingo, **60**
Flannery, T., 20, 27, 37, 106, 109, 120
Flat Rocks, **116**, 117, 248
flightlessness (birds), 92
Flinders Ranges, 203, **206**, 214, 216, **217**, 220, **221**
Flinders University, 261–2
Fordyce, E., 152
Fortey, R., **5**
fossil analysis, computer techniques, 30
Fractofusus, **214**
freshwater fish, 144
Friedrich, C.D., 129
Friends of the National Museum, 110
From Small Things Big Things Grow, 67

Galapagos Islands, 16, 163
Gallimimus bullatus, 120
gastroliths, **51**, 52
Gehling, J., 21, 203, 205, **206**, 214, 216, 220, 260
Gelt, D., 203, 290
Genyornis newtoni, 80, 81, 84, 85, **87**, 88–9, **90**, 91, 92, **93**, **94–5**, 96–7, 166, 168, **168**, **170–1**, **258–9**
giant salamander, 106, 138
ginkgo, 144
Glenie, R., 106
Glossopteris, 50
goanna, 80, 85
Gogo, 97–8
Gogo nodule, **97**
'Gogo reef reconstruction', 99, **102**

Gomez, E., 183
Gondwana, 97, 188, 193, 250
'Gondwanan Triassic fauna', **55**
Gould, J., 34
Grazhdankin, D., **201**
Great Barrier Reef, 99
'Great Russian Dinosaurs', 21, 82–3, 120, 200–2, 256
Greening of Gondwana, 81
Gregory, J.W., 162–3
Grey, K., 203
Griphognathus whitei, **98**, 99, **102**
gypaetines, 163, 164

hadrosaurs, 116
Haeckel, E., 37
Haig, D., 285
Hall, B., **70**, 86, 88
Hambleton, R., 257
Hambleton Ruff group, 257
Hamilton, K., 202, 257
Happy Birthday Old Timer, **288**
Hasegawa, Y., 202
Hecht, M., 86
Heezen, B., 17
'Hell's Dinos', **289**
herbivorous dinosaurs, 133
Herman, J., 110
Hernianax papuensis, **141**
hibernation, 123
Hildebrand, M., 163
Hillary, E., 78
Hine, K., 25
Hoban, M., 260–1
Hoffmann, C., **204**, 214
holdfast, 219
Hunt, N., 206

ichthyosaur, **292**, **296**
'Ilbandornis' lawsoni, 166
ilkaite, 108
insects, 138, 144
Institute of Vertebrate Paleontology and Paleoanthropology (IVPP), 47–8, 64–5, 67, 71, 234
International Geological Correlation Program (IGCP), 493, 205, 260
Inverloch, 106, **108**
Iron Curtain, 200
Ivantsov, A., **213**, 230

Japan, 64, 106, 138, 206
'Japanese Giant Salamander', **139**, **140**
jellyfish, 225
Jensen, S., 203
Jurassic, **180**, 194
Jurassic Park, 83, 193, 256, 295

Kadimakara, 52, 168
kangaroo, short-faced, *see Procoptodon goliah*
Keep, M., 285

Kielan-Jaworowska, Z., 242
Kimberella, 205
kiwis, 148, 151
　　kiwi feather clothing, 154
Klein, C., 203
Knight, F., 52, 168
Komarower, P., 203
Komodo Dragon, 80, 85, 88–9
Kool, G., **116**, 260
Kool, L., 106, **116**, 117, 260–1
Koolasuchus cleelandii, 106, 108, **136–7**, 141, 144
Koonwarra, 138, 144
Kryoryctes cadburyi, 117
Kühne, W., 236, 242
Kyoto, 206, 261

La Brea tar pits, 163
La Colonia Formation, 188
LaFlamme, M., 227
lagerstätte, 144
Lake Callabonna (SA), **25**, 88, 92
Lake Eyre Basin, **162**
Lance Creek (Wyoming), 193
Lanus, D., 183, 186
Latimeria, 98
Latz, P., 172
Lawson, P., **162**
Leaellynosaura amicagraphica, 81, **83**, **104**, **112–3**, **114**, **115**, 114–16, **118–19**, 120–3, **124–5**, **127**, **130–1**, 132–3
Leahy, N., 261, 263
Leonov, M., 203
Le Pichon, 17
Leptoceratops gracilis, **vii**
Libby, W., 15
Lightning Ridge, 243
lithographic drawing, 38
Lithops, **228**
Logan, M., 201–2
Long. J., 20, 98, 106, 274
Los Altares (Argentina), **183**
lungfish, 99, 133, **143**, 144
Luo, Z., 242
Lystrosaurus, **44**, **50**, **55**

MacIntosh, J., 186
Mackenzie Mountains, 203
Magnificent Mihirungs, 169, 175, 177, 261
magpie geese, 148, **176**
Mallophaga, 149
Mamenchisaurus hochuanensis, **64**, 65, **66**, 66–7, 72, **75**
Marino-Hadiwardoyo, 51
marsupials, 238, 251
mastodon, 64
Mayr, E., 164–5
McClellan, E., 66
McCoy gallery (National Museum of Victoria), 66–7, 88
McEvey, A., 18, 32–4

McKenna, M., 16, 165
McNamara, G., 26
megafauna, 85, 263–5, 274, 277
Megafauna: Facts and Fun for Students, 263
megafauna stamps, **263**, **271**, **277**,
Megalania prisca (Varanus priscus), **76**, 80, 81, 85–8, **87**, 89, **90**, 91–2, **93**, **94–5**, 96–7, **258–9**
Megalapteryx didinus, **146**, 149–53, **150**, 154, **155**, **157**, **159**
Megirian, D., 148, 168, 172
Meldrum, M., 33
Merritt, J., 8, 33
Mesodma, 188
Mesozoic fauna, 68, 183
Mesozoic mammals, 194
Messel, 58
Messelobunodon, **57**
micropalaeontology, 203
microscopic evidence, 133
mihirungs, 148, **165**
Miller, A., 163–4
Miller, L., 163
Minmi paravertebra, 120, **124–5**
Miocene, 54, **59**, **60**, **178–9**
'Miocene fauna of central Australia', **59**
missing data, 194–5, 250, 291
Mistaken Point, **10**, **223**
moas, 149–51, 153–4
mummified, 148, **151**, 153
　　See also *Megalapteryx didinus*
mollusc, 205
Molnar, R., 96, 106
Monash Science Centre, 20–1, **20**, 141, 153, 201–2, 256, 260–1, 263
Monash University, 19–21, 32, 34, 40, 65, 80, 82, 138, 150, 154, 200, 203, 256, 261, 262
Monet, C., 35
monitors, 85, 88
monkey puzzle tree, **192**
monotremes, 117, 235, **242**, **244**, 251
　　as 'living fossils', 250
Morton, S., 82, 203, 243
Moscow, 200, 206, 213
motorcycle rally (Rapid City), **289**
Mount Everest, 78
Murray, P., 148, 168–9, 172, 175, 261–2
multiple working hypotheses, 133
multituberculates, 188, 243
Museo Ciencias Bernardino Rividavia, 193
Museo Paleontologico Egidio Feruglio (MEF), 182–3, 185, 188, 193
Museum of Comparative Zoology (Berkeley), 86
Museum of New Zealand Te Papa Tongarewa, 149–50, 152
Museum Victoria, 18, 19, 20, 24, 32, 52, 64–65, 67, **70**, **75**, 82, 83, 86, 138, 200, 256, 262,
Muttaburrasaurus langdoni, 120, **121**, **122**, **123–4**

Nama Group, 209, **228**
Namibia, 203, 214, 216, **228**, **229**
Namibian Geological Survey, 230
Narbonne, G., 203, 214, 225, 227
National Geographic Society, 20, 188, 230
National Museum in Timor Leste, 80, 285
National Museum of New Zealand, *See* Museum of New Zealand Te Papa Tongarewa
National Museum of Victoria, *See* Museum Victoria
Nature, 243
naturalism in art, 58
nautiloid, **102–3**
Nautilus, 46
Nei Mongol (Inner Mongolia), 47
Neoceratodus, 98
neognathous, 148
Neophrontops, 163
Neoproterozoic, 207, 213, 261, 290
New South Wales, 243
New Zealand, 148, 150, 153–4
Newfoundland, **10**, 203, **215**, **223**, 225
Ngapakaldia, **59**
Nile, 184
non-avian dinosaurs, 116
Norris, R., 201
North American fossil vultures, 16
Northern Territory, 169
Northern Territory Museum, 82, 148

O Mundo Perdido Timor-Leste, 284–8, **286**, **287**
Old World vultures, *see* gypaetines
Oliver, W.R.B., 149
O'Neil, G., 257
Opdyke, N., 109
Ornimegalonyx, **61**
ornithomimosaur, 123
orthographic drawing, 43, 248
ostriches, 148, 151
Otago Museum (Dunedin), 152–3
Otway Basin, **112–13**, **132**
Otway beach, 290
Otway Ranges, 109–11, **116**
Ovenden, F., 78
Owen, R., 24, 37
oxbow lake, 111, 116

Pacific Science Explorer Expo, 260
palaelodids, **60**
palaeobotanical reconstruction, 126
Palaeocene, 54, **56**
palaeognathous, 148
palaeolatitude estimate, 109
palaomagnetic data, 109
palaeoniscoid, **142**
palaeoornithology, 256
Paleontological Institute (Russian Academy of Sciences), 21, 83, 138, 200–2
　　Precambrian Laboratory, 203, 205, **213**
panoramic scenes, 54, 270

parallax, 40, 249
pardalote, **6**
Paris Basin, 16
Parvancorina, **207**
Pascual, R., 188
Paso de Indios, 182–3
Patagonia, 196, 243
'Pat's prize locality', **183**
Pawley, K., 138
Pectinifrons abyssalis, **214**
penguins, 185
pennatulaceans, 219
permafrost, 108
Phanerozoic matgrounds, **231**
phenoxyethanol, 149
philatelic researchers, 256
Pholidocercus, **57**
phorusrhacoids, 148
photography, 38, 154, 156
photorealism, 96
Picasso, P., 38–9
Pickering, D., 261, 274
pigmentation, 92
Pittosporum, 96
placentals, 238, 240
Plane, M., 166
plastic flow, 250
plate tectonics, 164, **184**
Platypterigius australis, **292**, **296**
platypus, 18, 117, 235
Pleistocene, 54, **61**, 80, 84, 86
Pleistocene cave deposits, **262**
polar environment, 122
polar winter, 123
Prato, 206, 261
Precambrian matgrounds, **231**
prehistoric humans, 265
Procolophon, **55**
Procoptodon goliah, **254**, **258–9**, **262**, **266**, **267**, **268**
Protoplotus beauforti, 50, **51**, **53**
pseudotalonid, **236**, **237**
pseudotribosphenic teeth, 234, 235, **236**
Pteridinium, **204**, **228**
pterosaur, **124–5**
Puerta, P., 183

Qantas, 120, 122, 128, 182, 193, 202, 257, 291
Qantassaurus intrepidus, **128**, 182, 193, **252–3**, 291
Queen Victoria Museum and Art Gallery, 21, 82, 153, 200
Queensland Museum (Brisbane), 96
Quinkin, **165**

rails, 148
Rain, Steam and Speed, 35
Ramos-Horta, J., 284–5, **286**, 288
Rangea, **206**, **229**
rangeomorphs, **206**, **226**

digital line graphic, **227**
pectinate-shaped, **214**
spindle-shaped, **214**
Rapid City, **289**, 290
ratites, 148, 151
realism, 40, 42
realistic art, 35
Remington, F., 129
Renault, L., 286
retro-deformation, 249
reverse engineering, 86
rheas, 148
Rich, L., 20, 78, **115**, 138, 185
Rich, T., 20, 78
Ritchie, A., 98
Riversleigh, 166
RNA, 149, 151
Roberts, M., 263
rock breakers, 110
rock drills, 110
Romeo, E., 193
Rowley, M., 154
Royal Flying Doctor Service, 162
Rozanov, A., 21, 202
Ruff, N., 256
Runnegar, B., 203
Russia, 138, 200–1, 203, 214, 219, 225, 230, 257

San Joaquin Valley, 284
San Remo, 106
Sanderson (née Barton), N., 117, 240
Sanson, G., 32
Sarries, A., 188
sauropods, **180**, 185–6, 188, 194
Schaeffer, B, 165
Schaff, C., 243
Schopf, B., 203
Schroeder, N., **116**
Science, 243
scientific art, 73
sclerophyllous tree, 126
screamers, 169, **177**
sea-pens, 219
sea-squirts, 209
Second International Symposium on Gondwana Dinosaurs, 188
sedimentary structure, **108**
Seilacher, D., 230
Serendipaceratops arthurcclarkei, vii
Shark Bay, 49
shrimps (freshwater), 144
Shuotherium, 48, 50, 235, 238, 248
Shuotherium dongi, **234**, **237**, 238, 248
Shuotherium shilongi, **237**, 238
Siberia, 203, **209**
Sichuan, 186, 234
signalling behaviour, 92
Skinner, Morris & Mary, 165
Smith, D., 66, 78, 200
snake birds, 51

See also anhingas
Solenodon, **61**
South Africa, 206
South African Museum, 123
South America, 148, 169, **180**, 184
South Australia, 88, 203, **217**
South Australian Museum, 21, 80, 82, 168, 203, 260, 262
South Dakota School of Mines, 290
South Gippsland, 138
South Pole, 133
Southern Hemisphere, 242
Spielberg, S., 182, 193, 256, 295
Spriggina, **222**
Stein, G.W.H., 164
stereophotograph, 247
Steropodon galmani, 243
Stewart, I., 26
Stirling and Zietz expedition, 88–9, 168
Stirton, R.A., 15–16, 18-19, 162–4
stomach stones, *See* gastroliths
Stone, D., 200
stromatolites, 49
Strzelecki Group, 248
Strzelecki Ranges, 110, **116**
Stubbs, G., 37
submolariform, 242
Sumatra, 51–2
'Summer solstice ephemera 1', **141**
Sun, A., **65**
Suz'ma River, 207
swamp harriers, 32
Swinkles, P., 66
Sword Gusmao, K., 285
Sykes, L., 17
symmetry, 219
Symposium on Gondwana Dinosaurs, 182

Taibesi Market, 285
talonid, 235–6
tannin-stained water, 144
Tarka, C., 165
Tasmania, 278
Tasmanian Museum, 82
Tasmanian tigers, **82**, 277, **281**
Tassell, C., 21, 26, 153, 200
Tedford, R., 17–19, 163, 165
Tehuelchesaurus benitezii, **180**, **185**, **186**, **187**, 188, 193, 195, 196, 291
Teinolophos trusleri, **242**, 243, **244**, **245**, 248
temnospondyls, 106, **136–7**
Templeman, B., 81
tendon, 156
termite mound, **57**
Tetun, 285–6
Texas Tech University, 19
Tharp, M., 17
The Birds of Timor and Sumba, 164
The Fossil Book, 18, 46, 48–9, 50, 52, 54, 200, 261, 274
The Land Before Time, 193

The last, last labyrinthodont?, **136–7**
'The Last of the Mob', **263**, 278
The Lost World of Timor-Leste, 284–5, **286**, **287**
The Precambrian Biosphere: A Multidisciplinary Study, 203
The Rise of Animals. Evolution and Diversification of the Kingdom Animalia, 203, 260
three-dimensional imaging technique, 40
Thrinaxodon, **55**
thylacines, 277–8
Thylacinus cynocephalus, **258–9**, **281**
Thylacoleo carnifex, **258–9**, 261, **269**, **271**, **272**, **273**
Tianjin Museum, 64–5
Time magazine, 83, **120**, 257
Timimus hermani, **104**, 105, 120, 122–3
Timor-Leste, 257, 284–5
trace fossils, **230**
Traynor, M., 66
Trelew, 182, 185, 193
Trembath, J., 202
Triassic, 54, **55**
tribosphenic mammals, **232**, **237**
  molars, 235, **236**
Tribrachidium, **217**, 219, 290
Trigonid (teeth), 235, **236**, **237**
trilobites, 286
Trusler, G., 145
Tsintaosaurus spinorhinus, 65, 66, **70**, **71**, 72
tunicates, 209, 211

Turner, J.M.W., 35
turtles, 141, 144
Tyrannosaurus rex, **31**

University of California at Berkeley, 15–6, 86, 162–3
  Department of Paleontology, 16
University of California at Davis, 163
University of California – South Australian Museum expedition, **162**
University of Florida, 16, 109
Upland Moa, **146**, **152**, **155**, **157**, **159**
  See also *Megalapteryx didinus*

Vacca, R., 183–4
van Tets, G., 51, 168
varanid, 85
Varanus priscus, See *Megalania prisca*
Varanus komodoensis, 85
Vendian Period, 205, 214
Victorian coast, 126
vine forest, 169
visual distortion, 249–50
visual literacy, 74
volcanic ash, **181**, 196, 215, **276**, **277**
Vombatus ursinus, **29**

Walters, M., 203, 243
Wang, B., **65**
Wang, Y., 238
Warren, A., 138
Warren, J., 19, 65

watercolour painting, 40, 96
Waterhouse Club, 220
Wegener, A., 50, 164
Wellington Caves, 25
Wells, R., 17, 261–2, 274
Western Australian Museum, 98
Western cultural history, 11
White, M., 81
White Sea, 138, 200–1, 203–5, 207, **209**, **210**, 214, 216
Whitelaw, M., 109
Wildlife Art Society of Australasia, 29, 32–4
Wildlife of Gondwana, 80–3, 97, 126, 153, 169, 188, 201, 261, 263
Williams, C., 243
Wilson, B., 64–5
Wise, G., 145
wishbone, *see* furcula
Wolf, J., 37
wombat, **29**
Woodburne, M., 163
Worthy, T., 156, 274
Wyatt Durham, 16, 163

Yaldwin, J, 149
Yangtze River, 203
yinotheres, **234**
'Yorgia', **210**

Zhang, Y., 67
Zhou (Chow), M., 47–9, 65, 234, 236
Zhou, S., 67